PRAISE FOR
The TECH NERDS' Guide to CAREER SUCCESS

"*The Tech Nerds' Guide to Career Success* takes a fresh approach to helping engineers (and anyone who is serving customers on the ground) learn lessons that are important to them and to the company they work for. Dan's approach is to teach financial and business lessons in a powerful and understandable way. It allows people who are in roles delivering outcomes to connect the dots to the business side of what they do. Delivering exceptional technical outcomes is worthwhile, but far less so if disconnected from the other parts of the business, like strategy and operational excellence. This book is a must-read for people on every side of the table in a business. It provides valuable lessons and can be a platform for holding meaningful and crucial conversations to help everyone in a company understand how their work impacts their coworkers, customers, and the company they work for."

—**Arlin Sorensen**
Founder HTG/Evolve

"Many IT services leaders focus on their own understanding of their businesses, metrics, and operations to drive the business forward, yet wonder how to provide that insight to their staff. Dan's approach helps technical service people understand the 'why' of the service business in an easy-to-consume and understandable way. Told in parable, engineers get walked through the core tenets of a business and relate that knowledge right back to their work. Ever needed to help teach someone why documentation matters? This is your book."

—**Dave Sobel**
Host of the Business of Tech Podcast

"*The Tech Nerds' Guide to Career Success* is a fun way to teach Business 101 to people in any service business, but especially an IT managed service provider organization. It teaches through story. It's short and to the point, an easy read, and tees-up good discussions with your teams. You could even have a team act it out as improv with a narrator! Most techs and other MSP team members don't grasp these concepts. I recommend you test it with your team."

—**David G. Russell, CEO**
Manage 2 Win & Habitly

"Working with technical experts every day, I see many who have the potential to contribute to the business and grow in their careers but lack the professional skills to do it. This book shares how to shift thinking from a purely technical lens to providing value to the company and clients. Every chapter includes specific steps, so you know exactly what to do. It's as if you have your own personal coach. If you follow these lessons, you'll take your career to the next level and become sought after as either a foremost contributing business professional or a leader, should that be your desire.

—**Kendra Lee**
Bestselling author of *The Sales Magnet*

"Dan Adams has written the ultimate book to prepare technicians for success in managed services. No matter how good you are at technology, you need the stories and lessons in this book for your career and business success."

—**Don Crawley**
Author of *The Compassionate Geek*

"The mini-stories within the book illustrate what many business owners and executives are struggling to convey to valued team members. These stories will be useful to help take many great team members to the next level by teaching the game of business."

—**Steve Riat**
Industry speaker and Director of Sales at Nex-Tech

"This industry is full of brilliant people who have strong technical skills but haven't had the experience to unlock their potential as managers and/or leaders in the space.

This well-written fable will connect with those who desire to take the next step in moving from a job to a career in leadership. Well done."

—**Jamison West**
ConnectStrat – Strategic Coach
TimeZest – Cofounder and Chairman
MSP Talent – Industry Advisor

"TSP business owners are always asking 'How can I create a culture of accountability?' and 'How can I get my people to think more like an owner?' Dan has lived that reality and understands those gaps and how to address them. As a business grows, it needs people who understand how business can be conducted in a way that is a win for the company, for the client, and for themselves and their career. They just need to be shown what that looks like. The concepts in Dan's book are culture shaping. Get a few people to think and act differently, and you can experience a tipping point to transform your business into what you've envisioned."

—**Brad Schow**
VP of Business Transformation Evangelist
Growth at Connectwise

"In today's world, simply having a strong service team is no longer enough for MSPs to keep pace with their tech-savvy clients. To thrive in this rapidly evolving landscape, MSPs must also adopt a strategic approach, with a deep understanding of their clients' business needs, not just their technology requirements. But how can you make that happen? Dan shares his insightful experiences of a company that successfully goes through this transition. Each win and loss is dissected for valuable takeaways, culminating in a roadmap for meeting future challenges head-on. With this knowledge, you can confidently join the next generation of MSPs that stand out in the field."

—**Brian Doyle**
Cofounder & CEO
vCIOToolbox

"Dan Adams's book *The Tech Nerds' Guide to Career Success: Seven proven lessons to unlock your growth in an IT services company* is the missing piece to successfully growing your IT company. Most all business education resources are focused on the owner, and that is a great place to start, but if you want to grow, those business lessons need to make their way through to the rest of the team! Dan's book does just that, bringing powerful business lessons in an easy-to-read format. It is broken down into functional areas that provide context and that team members will be able to relate to. Your team will be able to understand not just the technical impact that they have in their actions, but the business impact as well!"

—**Shawn Walsh**
Partner, Encore Strategic

"*The Tech Nerds' Guide to Career Success* delivers great content for new techs into the MSP business that is not easily found elsewhere. The lessons learned are what differentiate the truly exceptional contributors from those who always struggle to figure out why they don't succeed. This should be required reading during onboarding to get staff started off right, from the start."

—**Rashaad Bajwa**
CEO, Integris

"Stop reading this review and put *The Tech Nerds' Guide to Career Success* in the hands of all your team members (not just techs) today—RIGHT NOW! And then congratulate yourself for taking the best step you have made in the past year to improve your business. As a former partner in Service Leadership, I can tell you Dan's book is a fun read and beautifully written, and it delivers easily understood, highly valuable lessons for all looking to succeed in business. WHY ARE YOU STILL READING THIS??? Put *The Tech Nerds Guide to Career Success* to work for you NOW; you won't regret it!

—**Brian O'Connell**
ChangePoint Advisors

The
TECH NERDS'
Guide to
CAREER SUCCESS

Seven proven lessons to unlock your growth in an IT services company

DAN ADAMS

Up-Skill LLC

The Tech Nerds' Guide to Career Success
Seven proven lessons to unlock your growth in an IT services company

Dan Adams

Published by Up-Skill LLC
Copyright © 2023 by Dan Adams
All rights reserved.

No part of this publication may be reproduced, distributed, or transmitted in any form or by any means, including photocopying, recording, or other electronic or mechanical methods, without the prior written permission of the publisher, except in the case of brief quotations embodied in critical reviews and certain other noncommercial uses permitted by copyright law. For permission requests, write to the publisher, addressed "Attention: Permissions Coordinator," at the address below.

Up-Skill LLC
Info@Up-Skill.com

Limit of Liability/Disclaimer of Warranty:

While the publisher and author have used their best efforts in preparing this book, they make no representations or warranties with respect to the accuracy or completeness of the contents of this book and specifically disclaim any implied warranties of merchantability or fitness for a particular purpose. No warranty may be created or extended by sales representatives or written sales materials. Neither the publisher nor author shall be liable for any loss of profit or any other commercial damages, including but not limited to special, incidental, consequential, or other damages.

Publishing and editorial team:
Author Bridge Media, www.AuthorBridgeMedia.com
Editor: Helen Chang
Publishing Manager: Laurie Aranda

Library of Congress Control Number: 2023918076

ISBN: 979-8-9880912-0-2 – softcover
ISBN: 979-8-9880912-1-9 – ebook
ISBN: 979-8-9880912-2-6 – audio

Ordering Information:

Quantity sales. Special discounts are available on quantity purchases by corporations, associations, and others. For details, contact the publisher at the address above.

Dedication

To those who know they have more potential and are willing to take that little step that makes all the difference.

Contents

Pre Note xiii
Introduction 1
Prologue 3
Chapter 1: The Right Game 13
Chapter 2: Daily Reps 28
Chapter 3: Napkins in Advance 41
Chapter 4: Mocha in Mocha 52
Chapter 5: The More-Expensive Option 67
Chapter 6: Catch and Juggle 84
Chapter 7: Sailing 99
Epilogue 113
Lesson summaries 116
Close 125
Acknowledgments 127
About the Author 129

Pre Note

A note for those outside of North America. In this book, I draw from metaphors from my two most beloved sports—which, unfortunately, are both called football. North American football and football, that beautiful game.

As I talk about both, and to reduce confusion, I cannot call them both football, so in full disclosure I call North American football *football* and then use a term heinous to many, *soccer*. I also call it a game instead of a match and a couple of other nonglobal terms. I apologize and mean no disrespect.

For those who might get frustrated with a clueless North American git calling your beloved sport soccer, may I ask that you take a moment, swear at me out loud to get it out of your system, and then, for a couple of chapters, look beyond the brief usage of the word and focus on the lesson that comes from that section. That lesson, when understood, brings power that grows careers and companies, and you deserve that.

Introduction

The technology and information technology (IT) industries are expanding faster than people can keep up, and companies need people with skills. You have technical ability and a desire to help. It's the perfect match—you can see yourself moving quickly forward on the tech highway to success.

You want to grow and aspire to raise the quality of life for you and yours, but after a few years it becomes clear that advancement does not "just happen," and truthfully, you have no clear idea how to advance reliably or predictably. You have strong technical skills, but no college class or industry technology training course gave you the understanding on how to move your career forward.

In this parable, we follow Marty—a very capable tech nerd, if you will, who has a high work ethic and strong desire to contribute. But his career is stalled, and he's frustrated. By chance, he meets a tenured professional services veteran who unlocks some "not so common" knowledge. These insights allow Marty to remove some career-limiting behaviors and turn them into advantages that magnify his impact in the business. As a result, he improves his career and life trajectory.

Everything you need to move your IT career forward and create the life you want is available to you. You just need to know where the levers are and how they work.

Prologue

Marty Young nervously twirled his pen between his thumb and forefinger. A tall, tan-skinned man, he had been at Tech Serve Co. (TSC) long enough that distinguished silver lines streaked his black hair and goatee. He and his engineering coworkers were crammed around a conference table that comfortably held twelve but currently had sixteen black mesh chairs stuffed around it.

Marty looked forward, eyes glued on the whiteboard at the front of the room, but he wasn't actually listening to the weekly engineering meeting. Only one thing mattered: Steve was about to announce the formalization of the new engineering manager position.

Steve was the president, owner, and CEO of TSC. He was a tall, bald man who always wore the company polo and tan slacks. He was just wrapping up the weekly meeting, and Marty knew an announcement was imminent.

"It's so exciting to have reached this point," Steve said. "We just hired our twelfth engineer last month! The team is getting too big for me to manage on my own. You've probably heard the rumors, and it's true: I'm creating an engineering manager position to give our engineers more direct guidance and leadership and help improve our results."

Marty grinned and twirled his pen around his fingers. He was sure that Steve would take him aside after the meeting and offer him the position. This was his opportunity for career success. He was arguably their strongest engineer, so much so that his coworkers often came to him with their tech questions. He had been putting in the time, leading projects regularly. He was ready for this. He deserved it.

I already know exactly what I'm going to do with that raise: First, I'll screen in my patio in the backyard. I'll get a big grill, maybe even put in a TV. I can work out there in the mornings. Then a new kitchen for Leia. And DJ and Linzey's college fund will be taken care of. No more stress about the future.

"I know there will be a bit of an adjustment period—you've been reporting to me, and I'm sorry I haven't had as much time as I'd have liked to support you. But this position will be in charge from here on out—I think you'll be really pleased to hear that Alan will be our new engineering manager."

The huge smile froze on Marty's face. *Alan? Did he just say he's already chosen the new manager? And it's Alan?* Marty looked around the room, struggling to process the revelation. Everyone was clapping, and belatedly he dropped his pen and added to the noise. Alan stood up, laughing and bashful, to accept Steve's hearty handshake.

Alan was saying something to the team, but Marty couldn't hear a word. He couldn't believe it. *Alan? Steve picked* Alan *over me? I deserved that promotion! I'm probably the best engineer they've got. What the hell does Alan have that I don't?*

Reeling, he tried to bring his attention back to the table. Alan was talking about communication and standards, saying a whole lot of words that amounted to a whole lot of nothing. Marty dragged on the neck of his polo. He could barely breathe. It was gone, just like that. All his plans. His eyes drifted back to the paper in front of him. He had been doodling plans for the patio, with a built-in grill and serving station and a flat-panel TV on the wall.

As the meeting adjourned, Marty scratched a savage line through the silly drawing. That's all it was, now. A stupid idea.

Marty left work ten minutes early so he could beat the traffic to his son DJ's soccer game. He played every Friday on the U-12 traveling team. Marty tried to never miss a game, but tech support work made his schedule unpredictable at times.

When he arrived, Marty saw that DJ's team was wearing its blue home jerseys, while the visitors were wearing orange. He looked for an open spot on the sideline, where parents were lined up pretty much shoulder to shoulder. Finding a spot, he began setting up his folding camp chair. Coolers were everywhere, and nearly every parent had a Yeti or coffee in hand. A lot of parents were on cell phones, doing work or playing solitaire and the like, rather than watching the pitch.

The warmer-than-usual spring had caused the mosquito population to swell from nasty to unbearable, and they were ravenous that evening. One prepared mother was advertising that she had extra repellent spray for those not wanting to donate blood. With all the local business sponsors' banners on the fence, Marty was surprised one of them didn't read *OFF! Mosquito Repellent.*

As the game started, Marty had already been bitten one too many times and decided to get some spray. As he reached the woman with the repellant, an older man with thick silver hair and startling green eyes let out a booming, warm laugh that pulled away Marty's attention.

As Marty held his breath and sprayed himself down, he couldn't help but listen to the conversation.

"Tony, I'm telling you, I knew you were getting it," the man said. "You made the changes and made the promotion happen for yourself. It's just the first step; the opportunities are going to explode. This is just the beginning for you."

Marty sighed. *Is everyone out there getting the promotions that keep passing me by?*

"What?" the man asked, slapping his leg in delight.

Prologue

"Twenty percent *and* an increased bonus structure? Well, you really nailed it. I tell you, I knew you had it in you."

Marty had finished dousing himself in the smelly repellant but decided to apply extra to his arms and neck to listen in—he wanted to hear where this conversation went.

Despite his age, the other man was wearing a company shirt not unlike Marty's own. He chuckled at something the other person said. "No, no, I appreciate you saying it was me, but I'm telling you, you did the work. I just laid out the lessons you needed to hear so that you could grow. Oh, hey, I'm at my grandson's soccer game. I've got to go. Yeah, see you tomorrow."

The man hung up, and Marty returned the can to the woman with a thank you, then walked away as inconspicuously as he could. As his mind drifted over what he had overheard, he returned to his seat. He felt a little guilty for listening in, but . . . a twenty percent raise? What kind of lessons was this guy teaching that had that kind of impact?

That's what I really need. Someone to show me how to really improve my game.

The game was a low-scoring affair. The lack of excitement in the stands was the kind of thing that forced parents to consider adding a little extra into their Yetis to keep them entertained.

That said, Marty did not notice much of the game because he spent most of the time lost in his own problems.

That distraction ended when the ball was kicked out on the sideline right near a row of parents. The "action" caused a buzz of excitement that woke the parents and caused a few to hold their drinks more tightly.

One player yelled, "Chuck!" and a stout, redheaded defender with a flattop buzzcut ran and picked up the ball to take the throw-in.

DJ, a forward on the blue team, streaked down the side as Chuck threw the ball. The toss was on point and led the striker and ball past the opposing defender on a clean breakaway—just the goalie to beat.

"GO, DJ!" Marty yelled, a proud parent.

Tweeeeeet!

Everyone's eyes turned to the ref, a fifteen-year-old kid in a striped jersey two sizes too big for him. The ref mimed throwing the ball over his head and lifted his foot and then pointed in the other direction.

One of the parents stood up to see better. "What happened? We had a breakaway! What's wrong?"

The redhaired defenseman who had tossed the ball hung his head. Chuck had lifted his foot on the throw-in. It had all been for nothing.

The other team was given a throw-in, and the game continued. Within a minute or so, three tweets from the ref proclaimed that the game was over, ending in a one–one draw.

The players gathered around their coaches and lined up

to do the high five, "good game" ritual between the teams. As they did, the parents worked to gather up kids, fold chairs, and discuss dinner plans.

DJ took his turn in the line, then ran over to his dad. "Great game," Marty said.

"Thanks, Dad!" DJ took the water bottle that Marty held out and took a long drink. "Can you wait a little before we go? I want to work with Chuck a bit," DJ said, making the request into a declaration.

Marty laughed. "Okay, but not too long."

"Sure!" DJ grabbed his ball and yelled, "CHUCK!" to get the defender's attention. Then he turned back to Marty. "Dad, can you help? If you kick the balls back to us, we won't have to chase them down and we can get more practice in."

Marty nodded, turning so he faced the pitch. His back was now to the families migrating to their cars.

DJ and Chuck took turns doing a throw-in toward Marty, who kicked the ball back to them for each new throw. Chuck had great power but struggled keeping that back foot down. Both DJ and Marty did their best to encourage and coach Chuck. They took it slower, breaking it down more, and after a few minutes Chuck was making steady progress.

"Is that your son?" The voice startled Marty. He was having fun helping out and had forgotten other parents were still behind him. He turned to see just one other man left—the white-haired man who had been talking on his cell phone earlier.

Off Marty's nod, the man smiled approvingly. "You're doing something right with him. It's kind of him to help out. That's my grandson who can't seem to keep his feet down for a throw-in. Soccer is new to him. He mainly plays football and baseball."

"It is a unique movement that you don't see duplicated in other sports," Marty agreed, then offered his hand. "Marty Young."

His gesture was met with a friendly handshake. "Chip. Chip Hall."

"DAD!" DJ yelled.

Marty turned to see that he had two soccer balls at his feet, and the boys were waiting. He hurriedly kicked the balls back to the boys and yelled, "Guys, practice together for a few, okay?"

Marty hesitated to ask the question on the tip of his tongue. But he decided, since they were already talking, that it wouldn't seem too strange if he brought up what he had heard. "I couldn't help overhearing your conversation when you were on the phone earlier—I was curious, if I may, what line of work are you in?"

"Oh, I work at GES—Gulf Energy Services." Chip pointed at the emerald-green logo on his crisp white button-up shirt. "We do engineering consulting for the energy services sector."

"Ah, nice. I'm in engineering too," Marty said. "Well, IT services engineering. Probably pretty different in your industry."

Chip chuckled. "Actually, probably not. I find engineering's pretty much engineering across the board, and especially on the business side."

"You sounded like you were talking mentorship," Marty said. "What do you do at GES, if you don't mind me asking?"

"That's actually a funny story," Chip said. "I used to be the VP of Professional Services there, but now I have a . . . well, unique role." He shielded his eyes from the setting sun as he watched his grandson practice keeping both feet down and throwing in the ball.

Marty laughed. "Consider me intrigued."

"I act as a coach and mentor to help engineers reach more of their potential, if you will, so both they and the company benefit."

"Wow. That sounds . . . really incredible."

"It's a pretty fantastic job," Chip agreed. "There's nothing quite like helping someone realize and unlock their potential for growth. Seeing what they're really capable of, you know? Most people have no idea how much of an impact they can make and how close greater success really is."

Marty hesitated, but he knew help only came to those who asked for it. What did he have to lose? "Listen, I'm sorry if this is an imposition, but I sure could use some of Tony's luck. It sounds like you've been doing this for a while, and I'm at a spot where I'm in over my head. I'd love to buy you a cup of coffee and pick your brain."

Chip smiled. "I like the values you taught your son. So, sure. Do you come to all of your son's games?" When Marty nodded, Chip said, "Perfect. Why don't we go for a walk after the game next week?"

"That sounds great." Pleased, Marty took a deep breath and slowly released it and felt some stress melt away. He thought he might see a path forward, finally.

When the next week came, it turned out that Marty had something he really wanted to talk to Chip about—and not necessarily what he had expected.

Chapter 1

THE RIGHT GAME

At the end of day Friday, Marty shut down his computer so he could leave to reach DJ's game on time. As he shoved the last item into his overstressed backpack, Taylor, a fellow engineer, appeared next to him with a cable, a ladder, and a mischievous smile.

"Yo, Marty," she said. "Before you head out, can I ask a favor? I need to get this cable run above the ceiling tile, and even with this ladder I'm not tall enough. But with you on the ladder, we—or you—could get it done."

Marty shook his head and playfully rolled his eyes. He put his backpack down on his chair and grabbed the ladder. "Sure. Let's do this."

Marty had just finished dropping the ceiling tile properly into place and avoiding the debris that always managed to get in his eyes when he had the feeling that someone was looking at him.

"Marty? Um, could we talk?" a voice stuttered. Marty looked down and saw Kunal standing below him.

Kunal's black hair was tied back in his signature ponytail, and he was wearing tan pants, blue slip-on Vans, and a button-down, short-sleeved shirt with his t-shirt showing through.

"Of course. What's up?" Marty stepped down from the ladder.

"It's about the onboarding project for Firefly," Kunal said. Marty was leading the Firefly project, with Kunal under him. "We've run into—well, there's a bit of a snag, and I'm not sure what to do."

"Okay. What are we looking at?" Marty smiled, trying to put his coworker at ease.

Kunal shifted his eyes away, his hands in constant motion. "Well, there's a bit of a discrepancy between what the client feels was included in the onboarding project and, well . . . reality."

"What do you mean?" Marty asked, trying to keep his voice level even as his stomach started to churn.

"They think the data migration was part of the project, but it is not."

"No, it definitely isn't," Marty agreed. "Onboarding is just onboarding. Are you sure they think it is?"

"Yeah, I was talking to the point of contact at Firefly," said Kunal. "She mentioned it sure will be nice when the onboarding is done and their data isn't on the old servers. She talked about not having to deal with the old infrastructure anymore and how great it will be with everything being in the cloud.

"That didn't sound right," Kunal continued, "so I probed a little to see if she was talking about a new project they wanted done. She looked at me like she was wondering what was wrong with me and said no, that it is not something different. That it was part of the onboarding project. She said that was one of the *main reasons* they signed on with us. Marty, I'm telling ya, they believe it is part of onboarding."

"Dammit." Marty tapped his foot, considering. "Do you have any idea why she thinks that? I was there, and I don't remember anything we said that would have given her that impression."

"Well . . ." Kunal hesitated. "I asked whom she talked to about the migration . . . and she said, um, you."

Marty blinked, too stunned for anything more. "What? I remember we talked about how it was something that needed to be done right away, but I never said it was part of the onboarding scope! This is crazy! What the . . . how . . . well . . . hmm. Kunal, you have seen their servers. How much time do you think it will take to do the migration too?"

Kunal wobbled his head, debating. "What they have isn't *that* big and scary, but it'll take some time, no question. Probably another thirty hours or so?"

"Remind me, what was the original scoped time for the project?" Marty asked.

"Just under four weeks of labor—about a hundred and fifty hours, I think."

Marty winced and gave a resigned shake of the head. "Well," he shrugged, "they're a new client, and we want to impress them. We've gotta do what we've gotta do to make the client happy." Marty glanced at his watch—he really needed to get on the road. "Go ahead and get that extra work done, but don't let the scope creep anymore."

Kunal's shoulders relaxed, and his hands finally fell still. "Yeah, I won't. Thanks, Marty."

"No problem," Marty said. Kunal turned and headed back to his desk, and Marty put the ladder away in a nearby closet.

As he closed the door, Marty jumped in surprise. Steve was standing close, waiting to get his attention, but Marty hadn't heard him approach. Steve's lips were pressed into a thin line, and Marty had to stop himself from pulling out his phone to look at the time. He was already pushing it not to be late for the game. *What now?*

"Did I just hear that there's some serious scope creep and we're going to be thirty hours over on the Firefly project?" Steve's arms were crossed, his words short and clipped.

"Yeah, the client said they were sure that was part of the project—even mentioned it was why they went with us. We know how to do the work—it's just going to take a little longer for us than we thought," Marty said.

Steve rubbed his forehead, sounding more exhausted than angry. "Do you know how much that extra time will impact our margin and change overall performance? That extra time cuts right into our margin. We cannot afford to give value away, Marty."

Marty winced. He hadn't looked at it like that. "I'm not sure on all that, and I'm sorry." Marty looked at the floor. "It just seemed like we don't have much of a choice here. They're a new client. And great customer service is one of our core values. We don't want to start out on a bad foot with them. I got a good sense when I was out with them, and I really believe they'll be a great reference client down the road. And I bet they upgrade to our highest managed service level by the next contract."

Steve grimaced. He looked at his watch, debating, then shook his head. "I know you have your kid's game, and I don't really have time to get into it now. Let's talk on Monday."

Great. And I'll have this hanging over my head all weekend, Marty thought. "Sure, yeah. Thanks, Steve. Sorry about that."

Steve just nodded and walked away.

Marty shoved his hands into his pockets, frustrated. He could see why Steve was upset about the project going over, but how much control did he think Marty had? The client was the one that had it all wrong. His call had just been whether to hurt the relationship right off the bat or go over on their hours.

The company wanted to make the client happy. Even if this did eat into their margin, it wasn't like they weren't getting paid for the work. It would just be a little less overall than Steve wanted. If they did a great job, the client would be happy. Happy clients paid their bills. It was the right decision.

Wasn't it?

The Houston rush hour traffic was as brutal as usual, and Marty was forced to exit I-69 for side streets to make any progress. When he arrived, he took off the cap he left in the truck and ran a hand through his hair to get rid of the ring the cap left behind. His wife, Leia, was already there, wearing her blue scrubs with yellow ducks on them and white Crocs. Their daughter, Linzey, was sitting beside her, reading a book, oblivious to the world.

Chip, the career coach Marty had met the week before, happened to be sitting just to the right of Linzey. He was dressed the same—in his company shirt, his silver hair a little wind-tossed. He smiled as Marty came over.

"Another week, another game," Chip quipped.

"Just glad I made it—traffic was unreal," Marty told him with a smile. He turned to his wife and gave her a quick peck of affection. "Sorry I'm a little late. How was your day?"

"Good, not too busy. DJ almost scored a goal. Great shot, just sailed a little too high. A little lower and it would have been perfect. How was work?"

Marty rolled his eyes. "You know." He turned his attention to Linzey as he set up his chair, but when he asked her about her day she just mumbled and kept her attention on her book. He chuckled and kindly patted the top of her head. At least she was reading and not scanning TikTok or playing on her phone.

Once he was settled, he pulled out a bottle of flat Diet Coke, which he habitually kept in his backpack and turned his attention to the game.

It was more engaging than last week's game. Both teams were moving the ball, with a lot of shots on goal and saves. So many close calls. Toward the end of the second half, the two teams found themselves drawn at two–two.

The opposing team's goalie had just made a great save, and everyone was backing up as he positioned for a goalie kick to start a drive.

But instead of kicking it, he threw the ball down the side to a defender, who continued moving it up the side. The defender sent a great pass to a forward on the same side. As the forward dribbled toward the goal, the defenders narrowed in on him. He passed it back a bit to a midfielder who was a good thirty yards from the goal.

Apparently, he saw an opportunity and surprised

everyone as he smashed a shot on goal further away than a shot would normally be taken.

A defender in the direct path of the ball turned around just in time to see the ball careening toward his head. Maybe it was self-preservation—or he forgot what game he was playing—but it didn't matter. He raised his hands and blocked the ball before it hit him in the face.

The crowd of parents and other family members collectively groaned as the ball connected with the defender's hands.

Tweeeet went the ref's whistle.

"And he's in the penalty box," Chip said, shaking his head.

Marty nodded. That would mean a penalty kick for the opposing team. "That's not good."

Chip shrugged. "Just goes to show, if you forget what game you're playing, even for a moment, the whole outcome can change."

After the game, Marty congratulated DJ on his play and the near miss on the goal, and he kissed Leia goodbye. He had previously told her he was going to try and connect with Chip.

Chip and his wife were just finishing folding up their chairs when Marty approached.

"Chip, do you still have time for that walk? I really need to bend your ear."

Chip glanced at his wife. She gave him a nod and knowing look that said, *I'll see you later.*

Chip smiled at her and turned to Marty. "Sure—let's walk the pitch."

Marty wanted to jump right into his fears about work, but he decided to be patient. As the two started to walk, Marty fell into his usual fast clip, but he noticed Chip's leisurely pace and had to deliberately slow down to match.

Evening was approaching, but the sun had not yet given up to the deep blue sky's pressure, producing brilliant orange, red, and pink hues. A warm breeze was picking up. On a sustained warm gust, Chip stopped, closed his eyes, and breathed it in with reverence. "It's my favorite time of the day. Granted, we're not on the water, but with a good sunset and a little breeze, I can almost smell the saltwater."

He exhaled, smiled, and turned to Marty. "Okay—you've been patient enough with this old man. What's up?"

Marty grinned and launched into it. "We have a new client onboarding project, and there was a scope misinterpretation. We had the choice to start off by confronting the client and hurting the relationship or do the work and make them happy. It wasn't a huge amount of time, and I figured it was pretty much a no-brainer—you put the time in and go over, right?"

"Hmm." Chip looked thoughtful. "I'm afraid I don't have enough information. Can I ask a question or two?"

"Fire away," Marty said, willing to share whatever he could.

Chip slipped his hands into the pockets of his khakis with his thumbs out on top. "Why is this a concern for you right now? What brought this up? You sound like you already know the right thing to do, so something has to have happened to make you question that and talk to me about it."

"My boss overheard me when I told an engineer to just put more time into the project."

Chip chuckled. "Oh, I see. Continue. What else did the 'boss man' say?"

"Let me try and get his words right," Marty said, eyes narrowing in thought. Unconsciously, he mimicked Steve's voice pattern. "'Do you know how much that is going to impact our margin and change overall performance . . . ?' And then something about giving value away."

Chip let out another soft *hmm*, bowed his head a little, and stopped walking. He spoke, still looking down. "Well, Marty—there's a bit to unpack there, but one thing is for sure . . ." He paused and looked directly at Marty. "You and Steve are speaking two different languages. You're talking tech and service, and he's talking business."

"Well, we have different areas of responsibility, that's true. I'm an engineer, and he's all business. But the right thing to do is 'do a great job, make the client happy.' After all, we are an IT services company, right?"

"Let me ask you this," Chip said. "What happens if you deliver great customer service, you're technically brilliant, and your customers love you, but you don't clear enough money to pay your business obligations, salaries, benefits, partners—you name it?"

Marty was taken aback. "What? That wouldn't happen, would it? I mean, they are still paying for the onboarding project."

"You can be great technically and have the business fail. I've seen it," Chip said, gesturing to highlight his point. "Great engineering does not equal success. Your clients can love you, and again, the business can fail. I've seen that happen. At the end of the day, Marty, you're in the game of business. The health of the business is a real indicator of doing it right. After all, why is the business doing this? Spending money, risking salaries, overhead, effort? What's the goal of all of that?"

Marty didn't answer, deep in thought, his head spinning a little.

"I know you work for a technical services company," Chip continued, "so technology is important. But it isn't the end goal. Customer service also is important—how we treat people really matters—but the biggest reason for the business to exist is to create margin. That, for the most part, is how the business scores. Money left over after all the obligations are met."

Marty shoved his hands in his pockets, a furrow

between his eyes. "I hear that, Chip, but that's a business problem. I'm tech, a service person. My focus is to be really good technically and make the client happy."

Chip just shook his head and looked at Marty. He paused, conviction in his green eyes.

"What?" Marty asked, a little uncomfortable.

"One word, Marty." Chip paused to make sure they locked eyes before he continued. "*Handball.*"

Marty's eyebrows rose.

Chip asked, "What was the impact of that handball on the game today?"

"It completely changed the outcome," Marty said.

Chip nodded. "All because, for one moment, a player either didn't know, or forgot, or deliberately chose not to follow a rule—and *BAM!*" Chip slapped his fist into his open palm.

"Let's consider that for a minute," he continued. "It's a ninety-minute game with two teams of eleven players. That's about two thousand minutes of game time—the same as about four eight-hour days. All it took was one person's mistake to impact the result of *four days* of effort."

Marty shook his head, thinking, as Chip continued. "Marty, let me frame it this way and ask you a question. For a team to win, who on the team needs to know the rules to be successful? Just the coach? A couple of the key players?"

"Well, everyone on the field, I'd imagine," Marty answered.

"Why?" Asked Chip.

"Because everyone has the ability to change the outcome," Marty answered.

"Exactly!" Chip raised both arms up as if giving a hallelujah in church. "Every player on the field needs to know the rules, or the chances of winning drop exponentially." He paused, putting emphasis on the next four words. "*Business. Is. The. Same.* Marty, there are rules to business: ones that cause penalties as well as ones you need to know if you want to score."

Chip tucked his hands into his pockets. "If you want success, you need to know how to optimize the odds of winning. And make no mistake about it, everyone's efforts directly impact results. Everyone needs to know what the rules are and what success is, or the team will *not* succeed."

"Well, when you put it like that . . ." Marty laughed sheepishly.

"Seems obvious, right?" Chip grinned, glad to see Marty taking it all in. "And one other small insight here, if I may, Marty? I know this one can feel a little cold: How do coaches and even teammates feel about someone who keeps 'costing them the game,' so to speak?" He paused again before continuing. "Coaches want players who add value and don't cause losses. And they want—and need—great players who know how to make everyone better."

Marty sighed. "I really put my foot in my mouth, didn't

I? I was so sure we just had to be great technically and make the customer happy."

"First, let me say you're being strong to open up to this," Chip assured him. "It hurts to learn that actions with good intentions might not have been doing as well as you believed they were. Growth can be painful. Most of us in technology or service"—he pointed to the GES logo on his company polo—"know our craft, and we want to help people. But nobody pointed out the rules to the bigger game we're playing. Concepts like utilization, standards, business models, and margins."

"Yeah." Marty laughed. "I wasn't totally sure what Steve even meant by all that."

"Business words like *margin* can sound like bad words, but they're not," Chip said. "They're just lingo tied to the game of business. Like knowing what 'offside' means or what a 'throw-in' is. They're fundamental, essential. When you know what they are and how they impact the game of business, you can reduce game-changing mistakes and amplify the actions that help win."

"So . . . those business fundamentals, those rules of the game. How do I learn those?"

Chip grinned. "Well, asking that question is a great first step."

Marty nodded, torn between the feeling of disappointment in his predicament and a welling desire to fix it. He knew he was a good troubleshooter. This was just a different set of problems—and he was starting to understand the equation.

He was ready for the lessons Chip had to teach. Marty was just getting started.

Marty spent the weekend waffling between anxiety and exhilaration. Chip had sent him some training to start developing a working knowledge of key service business fundamentals. But he also had the threat of Steve's "talk" hanging over his head. Would Marty be able to bring what Chip had taught him into practice? Or would Steve not want to listen?

Lesson 1: You're in the game of business.

Chapter 2

DAILY REPS

On Monday, Marty came into the office full of nervous energy. He went to his desk, shoved his bulging backpack at his feet, and checked his email. Sure enough, he found a message from Steve asking him to come to his office at ten, which meant an hour and a half of waiting. Marty knew Steve wasn't happy about the overage of the onboarding project; thanks to Chip's advice, Marty was ready to apologize for his call.

He caught up on emails, then worked for a while on a project plan that the account manager kept pestering him to finish. But he could not concentrate on it and failed to do any useful work.

Finally, the clock ticked closer to ten and the minutes that had been dragging started to race. Marty found himself opening files and looking for an excuse to be a few minutes late. But he knew that would look bad and reluctantly dragged himself out of his ergo chair to walk over to Steve's office.

Chapter 2: Daily Reps

Steve had the corner office with big windows looking out over a parking lot and another huge office building across the way. His standing desk was crammed with three monitors, a keyboard, a cup of coffee, and personal knickknacks. The office barely had room for a second, small, round table with three chairs around it. When Marty entered the office, he was surprised to see Alan, the engineering manager who had received the promotion, sitting at the table and looking awkward.

Steve was on his cell phone but put it down when Marty entered. "Thanks for coming in." He gestured to the last open chair.

Oh great, Marty thought. *It's an ambush.* He closed the door behind him, though the floor-to-ceiling windows still made him feel exposed. He considered, and then immediately tossed, the idea of small talk. He wanted to get right to it. "So, I guess we're talking Firefly," he said, taking his seat.

"Firefly is part of it," Steve acknowledged. "But we have some larger concerns with trends we're seeing that we want to work with you to improve. We need to talk about some performance concerns."

Marty felt his brain go blank. He had been prepared to counter their concerns over his Firefly onboarding call, bringing up everything Chip had talked to him about on Friday. But larger concerns? What did Steve mean?

"You have so much raw ability," Alan chimed in. "You're one of our strongest engineers by far. The problem is . . . we're just not seeing the consistent performance we need from you."

Steve smoothly took over. "Marty, Alan and I have prepared a PIP to help us move forward. Alan, would you please take us through it?"

Alan pulled out a two-page document and placed it on the table. It read *Performance Improvement Plan for Martin Young* at the top. Marty scanned it and saw bullet points of actions, broken into specific focus areas for granular performance concerns. The top three areas were project management, timeliness and accuracy of time entries, and effective internal communications.

Alan started talking, but Marty was already just keeping his head above water.

For the next ten minutes, Alan stepped through each area, with Steve adding commentary along the way. Marty felt like the ball in a ping-pong match—smacked back and forth by two players, both looking for every opportunity to spike the ball. He wasn't sure if they expected him to talk, but even if they did, he had no idea what to say.

He came to consciousness when Alan said, "I know you're working on these things, but you're not updating our management software to keep others informed as to what's going on. You're keeping too much inside your mind, and it's causing problems for others who try to work beside you." Alan made earnest eye contact, which Marty could only keep for a moment before looking away.

"Sometimes you're on top of it, and things go well, ," Alan continued. "But when you're not, others are forced to scramble to overcome the holes . We waste a lot of time and effort."

"That costs us a lot of money," Steve said, adding his volley. "For instance, that onboarding for Firefly. Not only are we spending extra time giving them something for free, which hurts the margin on that project, it hurts us again because we can't charge them for the value of the migration. It's a double negative."

They paused, long enough that Marty felt like they were waiting for something. He cleared his throat and looked up. "Guys, you know I'm working hard . . ."

"Marty, you're working hard. We're not questioning that." Alan spoke with real warmth in his voice. "It's putting effort into these little details that's essential here. Somehow you need to understand that these things really make the difference, because when they are missed, the cost to the business is significant."

Marty nodded, smoothing his fingers over the paper in front of him. He didn't have anything else to say.

Alan sighed and leaned back in his chair. The worst of it was over. "I know you've got it in you. I'll work with you on a regular basis to knock out this PIP and ensure you make these actions habit, and we can move this all forward. Instead of our monthly one on ones, we will meet every other week till we are past this."

"We love working with you," Steve finished up. "We really do want you here at TSC and part of the engineering team. But that said, Marty, if you can't make these adjustments, we will need to talk about your future with us."

Your future. Fired. Steve's words echoed in Marty's head. *That's the word he's talking around. If I don't get my act together . . .* Marty nodded. "I understand," he managed to croak.

"We all have areas we can improve in," Alan said, trying to bring the room back up emotionally. "I really believe you can make these changes and fix this, Marty."

"Thanks. I will." Marty stood up, clutching the papers hard enough that they creased in his hand. "I *am* dedicated to TSC," he promised, then quickly let himself out of the office.

Marty went back to his desk and slid into his chair, dropping the papers facedown so no one else would see them. A performance improvement plan. He'd never been called out at work like this. He had always done his work and done it well. *I kill myself! I work like a damn dog.*

He turned the papers over surreptitiously and rescanned the performance goals they had listed. Laid out in black and white like that . . . *They all seem doable. Most of this stuff isn't even technical. Hell, none of it is really technical. But if this is what they want from me . . . I don't know. I get so much done, but all they see are the times when I miss one of the things they want.*

Chapter 2: Daily Reps

Marty grabbed his cell phone, stuck the PIP into his backpack, and slipped into the office kitchen for some privacy. He dialed Chip and waited a moment while it connected.

Chip picked up on the second ring. "Marty! Good to hear from you."

"Hey, Chip. How's your day going?"

"Good, good. Yours?"

Marty laughed dryly. "Uh, funny you should ask. You mentioned that if things were a little hairy at work I could call."

"Absolutely."

"Well, I'm calling. I just had"—Marty blew out a long breath—"a bit of a *conversation* with my boss. I'd love to unlock some greater knowledge on this one. I'm a little confused, to say the least."

"Hmm. Sounds pressing." The line went silent as Chip checked his calendar. "I know I said we'd have coffee, but this week is a bit of a mess. How do you feel about ellipticals?"

On Wednesday morning, Marty pulled up in front of the gym and gave way to a huge yawn. He was a morning person, but getting up two hours early to hit the gym before work was a bit of a stretch, even for him.

Marty was greeted with that unique gym smell of rubber and cleaning solution as he signed in at the desk under Chip's guest account. The gym was midsized and well maintained, with a bank of ellipticals, steppers, and treadmills on one wall and a host of strength-building machines carefully placed around the different sections. Chip, wearing neon green sweatpants, was already in the adjacent stretching area when Marty walked in. *Damn,* Marty thought as he shook his head. *This guy may know business, but he has zero fashion sense.*

"Hello," called Chip when he saw Marty. "This old body needs a couple more minutes of stretching or it won't be pretty." Marty nodded and joined in.

When they were done, Chip asked, "What are you working on today?"

Marty laughed as he surveyed the gym. "You invited me," he reminded his mentor. "It's your call."

"True." Chip chuckled. "I'm working on lower body and some cardio."

"Good by me," Marty said, and they started over to the leg press machines. They took a couple of machines that were side by side. Marty had a small white gym towel thrown across his wide shoulders, and the workout shorts he was wearing showed off his white socks.

"So, trouble with the boss?" Chip pressed after the silence stretched out. He wiped a palm on his neon green track pants and moved the pin to adjust the weights.

"Yeah." Marty sighed. "They've put me on a performance improvement plan."

Chip winced in sympathy. "Ooh. Was that a surprise? What drove that?"

"I don't know. Consistency, I guess. It's like they don't see all of the stuff I'm doing. But then . . ." Marty's voice faded out as he focused on his reps. He counted each one out, making sure he was bringing his legs up to ninety degrees before he pushed back down.

"Consistency in what?" Chip prompted.

"Well, in the work I do."

Chip looked puzzled. "Help me understand. Can you clarify, please?"

"Well, sometimes I'm busy working on engineering items and don't do all the things they ask for," Marty explained as he started on his second set. "Finish the documentation, keep the project info up to date, enter my time in the way *they* want. I'm working hard; I'm helping as much as I can. I know they ask for this stuff, but I have more important things to focus on."

Marty finished his second set of squats on the machine. "Thanks for the details," Chip said, taking a break between his own reps. He nodded his chin at Marty's machine. "Let me ask you. Why did you make sure to get your squat to ninety degrees, thighs parallel to the platform, before starting back up? And why were you counting reps and then taking a break?"

Marty looked over at him, surprised. *Does Chip not know how to work out?* "Um, that's the proper form. It's the way you're supposed to do it to get the most benefit."

"Where did you learn to do it that way? Did you come

up with this on your own?" The glints in Chip's eyes sparkled.

"It's what the trainers preach, isn't it? Pretty much common knowledge at this point that form matters, as does breaking it down into sets of reps," Marty said.

"So let me get this straight," Chip said. "It's not something you came up with, and not 'the way' you might feel you should do it, but you listened to trainers about form, sets, whatever, to understand what to do in order to bring the most benefit?"

Marty took that in, and then laughed. "Hold on—this is a metaphor."

Chip laughed, the sound booming out across the busy gym, and a few people turned to see what was going on before returning to their workouts. "And a damn good one. Think about it. You're working hard, right? But if you're not doing the little things right, you're not getting all the value out of it."

Marty just shook his head.

"And while we're on metaphors," Chip continued, "here are a couple more that align for you. What happens if you hit the gym really hard but don't come back for a couple of weeks?"

"It wouldn't do much, other than make me really sore," Marty said.

Chip nodded. "Exactly. You have to be consistent to really get results. You can't binge success."

"That does makes sense." Marty wiped the beads of sweat off his top lip and started his final reps.

Chip started in on his third set. "The last concept that ties to this . . . Do you get big muscles and then go to the gym, or do you go to the gym and the muscles come from your hard work?"

"You have to do the work first, obviously."

"Just like at the gym, to get results at work, you have to put the effort in first." Chip was starting to breath hard from the exercise, but his voice was still clear as he said, "No one else can do it for you. You must work to unlock your own growth, consistently. Random workouts will make you nothing but sore. And you have to do it with the correct form. Anything different from this, you won't get the maximum results."

"I don't think I'll ever look at a gym the same way," Marty said with a laugh.

Chip's answering chuckle again was practically a sonic boom. "Just wait until we reach the treadmill," he joked. "I've got some thoughts on pacing there too. But before we hit the cardio to wrap up, you need to understand one more critical component that is relevant to your success."

"And what is that?" Marty stopped and tossed his towel over his shoulder.

"Oh, no you don't." Chip reached across and poked Marty in the shoulder. "No quitting on this one. Let's keep lifting—I'll share as we go."

After they set up on two adjacent leg extension machines, Chip started in on his lesson. "I have no doubt you're working hard and doing so consistently. But that said, you need to remember that service is all too often subjective."

"Subjective?"

"Yes, subjective. Since people deliver and receive the service, everyone can have their own interpretation of what success is. And do all people have the same knowledge? Understanding? Even want the same thing? No way." Chip scoffed. "It is tough, but you can be delivering the service to the best of your ability, doing what you believe success to be, but still fail by missing what others view as success."

Marty cocked his head and furrowed his brow. You could almost see and smell the smoke coming from his mind as it started to click. "As the person delivering the service, you can't stop at *your* interpretation of success. You have to align with the other person's interpretation."

Chip was breathing hard, though whether from the exercise machine or just passion at his message Marty wasn't sure.

Chip nodded. "You need to learn what *they* view as success—that's the real measuring stick. Funny thing, when this 'misalignment' problem shows up, it often manifests as a lack of consistency."

"A lack of consistency?" Marty asked, surprised.

"Think about it. Sometimes your interpretation aligns

with their targets and you meet their expectations, but sometimes you don't," Chip explained. "The way you see it, you're consistently working hard, just like you would want them to do."

"But on their side, they can't figure out why you're on and then off," Chip continued. "Sometimes you hit what they want, so they know you can do it. But then other times you don't, and they can't figure out why you're not consistent. It can be really frustrating for them. They start to wonder if you care. You really need to dial into *their* view of success—not yours."

"You're making me rethink a lot," Marty said. "Before this, I could never understand why they would even complain . . . but I guess now that makes sense. Crud, I don't know what's working me out more—the physical or mental exertion."

"Speaking of physical exertion," Chip said as they both stood up, "You must have worked up an appetite. Let's go out to lunch next time."

"Yes! I have to take you to Gas Station BBQ!" Marty grinned. "It's incredible. They converted an old gas station into a restaurant. Best pulled pork you'll ever eat, and the service is amazing. We should go there for our next mentoring session—my treat."

"I'm in," Chip said with his booming laugh.

Now Marty had some great barbecue to look forward to—and a lot of work to think about. He knew that he could make this idea count—he just had to clarify how those assigning the work viewed success and make that the target. And in doing so, he would nail his own success.

Lesson 2: Their interpretation of success matters, not yours.

Chapter 3

Napkins in Advance

Marty had his elbow out the window of his blue Toyota Tacoma, enjoying the wind moving through the cab, when his phone rang. He was on the way to the last client visit of the day, but he was used to taking calls on the go. As it happened so frequently, he had jokingly dubbed his truck his "home office."

He picked up through the truck's Bluetooth with a cheerful "Marty here!"

"Marty, this is Joe," a voice barked in a dark, gravelly snarl. Joe was the chief information officer, or CIO, of a company that TSC supported. Marty's coworker Kyle was doing infrastructure work at Joe's office this week.

"What the hell is going on with your guy, Marty? This Kyle is a real piece of work, let me tell you!"

He was talking so quickly it took Marty a second to figure out what was going on. "Sorry, Joe. What's happening?" Marty slowed his truck a bit, splitting his concentration between the call and the road.

"Did I stutter? What did I just say? That man of yours is completely incompetent, an absolute hazard! Our systems have been down *all day*, Marty! All day!"

"I'm so sorry," Marty said. "Joe, if I remember right, the work Kyle is doing is on the servers, switches, and firewall, and to get it done will require them to be rebooted several times. It's normal for them to be offline for a while."

"Oh, is that right? Do you think I know what tech pieces do what? If we were going to be impacted, do you think that would have been a good thing to tell us?" Joe snapped. "It's only our entire business we can't run!"

"Well, Joe, he should have let you know before he did it so your team could work around the outages. I am gathering he did not bring this up," Marty said timidly.

"He damn well didn't!" Joe's voice rose to an angry yell. "I've got angry emails, I've got angry phone calls, I'm taking damn hits here! And when I asked your guy about it, he just throws up his hands and says that it's normal, blah blah blah! Well, let me tell you, it sure isn't normal for none of our people to be able to work right before month-end! Does the guy have any damn clue what we do here at all? It's been a damn disaster!"

Joe was stringing together words so quickly that Marty could barely follow along, but the anger and frustration were clear.

"Joe, Joe, I am so sorry. That should not have happened.

He absolutely should have communicated to you that the work he's doing needs a system shutdown—"

Joe cut off the apology. "My company holds me responsible as the connection between you and them. If I don't know what is going on, I look like an idiot. I cannot be in the dark about the damn work you people are doing! Do you hear me?" Every word was a furious bullet in Marty's ear.

"Absolutely, Joe. I will talk to Kyle."

Joe was having none of it. "I don't want to see him here again!" Marty could practically hear the spit hitting the phone. "You pull him off our account *right now*, or we will walk away, I swear to God! Contract or not."

Marty winced. He really needed Kyle on Joe's account. Marty was slammed with other work. If Kyle wasn't allowed to do the work, Marty would have to shift his own schedule to fit it in. It would be a nightmare. "Of course, I hear you, and we will pull him off the project right now. You'll deal directly with me for everything in the future, okay? Does that work for you?"

"It's a start," Joe huffed. "That guy is completely incompetent. This is our month-end, Marty."

Marty needed a way to cool down Joe and make amends. "He should have known he couldn't take your systems offline without clearing the outage with you, and I'm sorry he didn't take that step. We can do this. I want you to feel confident in everything we're doing, and I'm going to figure out how to make sure this never happens again, okay?"

"Fine. Yeah."

"Okay, Joe, we will talk soon." Marty cringed as the line went dead. *What a headache! What was Kyle thinking?! Was he even thinking? And how the heck am I going to fit this into my schedule now?*

That evening Marty pulled in at the Gas Station BBQ and turned off his truck. He removed his cap, left it on the seat, and headed inside. The surroundings always put a smile on his face. Old license plates, interstate signs, and '50s gas station memorabilia plastered the walls. The smell of hickory smoke filled the room, causing his mouth to water.

A woman in a mechanics shirt covered in black stains came to seat him, her name tag proudly declaring her name as Bob. "Table for one?" she asked.

"Two, please," he said. He was meeting Chip here for a mentoring session. They'd both been busy over the past few weeks, but they had finally found time to meet and eat.

The waitress led him over to a blue-and-chrome booth with a view of large black smokers on the side of the building. Barely a minute later, a different waitress, also named Bob, led Chip over to the table. He was taking the place in with delight.

Chip slid into the booth, across from Marty, and they

Chapter 3: Napkins in Advance

chatted about family and life as they waited for the waitress to come back with menus. Chip coughed and patted his chest. He looked around the table, but the waitress had not yet brought them water.

"Excuse me!" Marty called, waiving a waitress down. "Can we get some water, please?"

"Of course you can, dear! Sorry, I thought I brought those already. " She hurried off and came back a moment later with two waters. "I'll be right back to take your order," she said as she handed them over.

"Um, we don't have menus yet," Marty reminded her gently.

"Oh my, no you don't! I am so sorry about that." She put a hand to her chest in alarm. "I'll be right back!" she said before vanishing.

She reappeared quickly with two menus.. "I'll give you two a moment, and I'll be back to get your order."

"Great could we order some drinks first?" Marty prompted before she could turn away. He tried to keep the frustration out of his voice, but the service was starting to get to him. Marty had promised Chip that this place had great service, but today was not going well. Was she going to do the basics, he wondered, let alone anticipate a single thing they needed?

She took their drink orders—a Jameson, ginger, and lime for Chip and a local hoppy microbrew for Marty—and hurried away. They watched her move to another table;

they could tell something was missing there as well, and she scampered away again.

"With as much as she is running around," Chip said, "she should be wearing a tracksuit, not a mechanic's shirt."

Marty laughed, nodding, and they both took a moment to look over the menu. When they were ready, Marty looked around. Their waitress was still hurrying to and fro, and he waved to indicate they were ready.

She returned, took a deep breath, and pulled out her order pad. "Okay, hon, what can I get you two?"

"I'll have the Oinking Convertible with Diesel," Marty said. All of the menu items had fun names, and the sauces they offered were unleaded, leaded, and diesel.

"I don't know these fancy names," said Chip. "I'd just like the salad with some smoked pulled chicken."

"Ah, a Skinny Clucker for you. I'd suggest leaded sauce on that," she cheerfully recommended.

"Sounds great." Chip held out his menu for her to take, wagging it slightly to get her attention. She grabbed the menus and disappeared back into the kitchen.

Marty and Chip's conversation drifted to the subject of work. But this time, they talked about Chip and his trainings.

Marty was firing off questions to better understand what was included. "How long does it take? Is there a graduation?"

"Well, . . . the training generally takes about six to

eighteen months to complete. At first it was very informal, nothing to mark the ending really." Chip was saying. "But after I had been doing it for a while, one of the guys joked that he used to walk around with a chip on his shoulder toward the business guys, but after we had worked together, he had a different 'Chip' on his shoulder." Chip made air quotes around his name, and Marty laughed. "One thing led to another, and now we have commemoration plaques with a 'Chip' on it."

"So everyone you mentor gets one of these plaques?" Marty asked.

"Yup, with a little chip of white marble from Thassos. Oh!" Chips eyes lit up. "I need to tell you about Thassos, Marty. It's an amazing Grecian island where the water and sailing are to die for. I was lining up retirement and spending more time there when GES asked me to stick around a few more years and really focus on the mentorship program."

The waitress came back with two plates piled high. The red-and-white checkered paper underneath each meal looked like the flag to start the Daytona 500. Marty rubbed his hands together as the waitress set each plate down.

"Y'all need anything else?" she asked.

"Um, I don't seem to have any utensils," Chip pointed out.

"Oh, snap. Let me grab that for you!" she said and turned to leave.

As she dashed off, Marty called after her, "And napkins! We're going to need lots of napkins!"

"Right!" She raised her hand to confirm she had heard. "Napkins! I'll grab those too."

Marty shook his head as she scurried off, lips pressed into a thin line. Here he was, trying to show off for Chip, and everything seemed to require prompting. "I don't know what's going on here today," he said by way of apology. "Usually 'Bob' is way better."

He waited to grab his sandwich, since Chip couldn't eat without his cutlery, but the sight of that food was almost enough to break his resolve. He took a moment to savor the smell, frustrated that he couldn't dig in, but he had been raised with too many good southern manners not to be polite.

Finally, the waitress returned with cutlery and napkins, and Marty was free to reach in and pull off a chunk of the pork bark. He just let the crust melt on his tongue like a chocolate truffle. The smoky, salty goodness sunk in, and he closed his eyes in delight. Across from him, Chip was equally carried away by his chicken, and they ate for a while in blissful silence.

Only when he had finished the sandwich did Marty settle back, taking a breather before attacking his coleslaw. "I'm glad I have your brain to pick. I've got an ugly one I can't wait to hear your thoughts on." Marty explained the problem he was having. "The client sees Kyle as incompetent, but the worst part is, he really is good at his job. He just—"

"Doesn't anticipate, think, and plan ahead, or prepare the client," Chip said, finishing Marty's sentence. "It's kinda like Bob, our waitress. She's technically proficient. She's friendly, she didn't spill anything, she got our orders right. But she didn't anticipate a single one of our needs. No water, no napkins, no cutlery. Heck, she didn't even bring menus with her! That leaves us wondering if she's competent. It doesn't inspire a lot of trust."

Marty wiped his chin with a napkin. "Huh. With Kyle I was thinking it was about poor communication, but you could be right. He didn't think ahead to consider the impact the outage would have on the business or who needed to know about it."

"Unfortunately," Chip said, shaking his head in frustration, "I see this *all* the time. I am constantly reminding engineers that clients can't measure your tech skills—they barely understand what you're doing. Hell, most of the time, that's why they hire you in the first place! But what they *can* tell, very clearly, are the actions that are the hallmarks of a real professional. For them, your competence is measured beyond technology."

"Hmm. That does sounds right," Marty agreed. "You pretty much need technical skill to assess technical skill. Without that, you really don't know what tech skill they really have." He finished off his coleslaw and put down his fork with a contented sigh.

"Exactly." Chip gestured with his fork like it was a

pointer. "Now, they may not know technology, but they sure know if you're late, if you don't make an effort to reach them or do what you said you would, if you speak in a condescending manner or don't keep them updated on progress. They know if you don't ask the right questions or keep good documentation. They know if you verify results and pay attention to details. All of those are attributes of a competent professional."

Chip dug into the salad and chewed before continuing. "People want competent professionals, not just a tech resource. Oh, and one area that really is sensitive with them is whether you're looking out for their best interest—protecting them and their business."

"What do you mean?" Marty asked.

"If you fail to warn them of potential impacts or do anything that makes them look bad, they let you have it real quick. But if you show them through your actions that you are aware and always keep them in the know, their trust in you skyrockets."

"That makes so much sense." Marty balled up his napkin and tossed it onto his plate. "Thinking back, a lot of the problems I've had were tied to failing to do *those* types of things versus lack of technical skill."

Chip put down his fork, chewing his last bite of salad, and folded his hands across his lap. "Your actions show that you know what's going on, it isn't your first rodeo, and you know what matters to them—that you are a real professional and they can trust you. You do that by communicating with them, keeping them in the loop and involved. Those anticipatory actions are hallmarks of competence. Your competence is measured in so much more than your technical skills."

Marty nodded. He could use this to help Kyle understand what went wrong, so hopefully this never happened again. Marty could improve his own habits in this area too. One of the items on his PIP—effective internal communications—tied directly to this problem.

From here on out, Marty would make sure Joe and all of his clients were informed and aware of what was going on and of any potential impacts of everything that he touched. No more letting these things destroy clients' trust. Marty took a big bite of his pickle, which he had saved for last, and smiled to himself. *This is going to be good.*

He was going to do the things that marked him as competent in the eyes of those he helped.

He could finally see the way forward. He had a vacation coming up with his family, and he needed to get a lot of ducks in a row first, making sure he had everything organized so the clients would be fine until his return. While he was excited to get away, part of him looked forward to getting back, refreshed, to continue his progress.

Lesson 3: Your competence is measured beyond your tech skills.

Chapter 4

MOCHA IN MOCHA

The first day back after a vacation always filled Marty with excitement and—truthfully—some dread. Although rested, he was nervous about his workload and anxious to see what new jobs might be on his plate.

This time was a little different, though. He had added courage: he felt that he was becoming more effective from the insights and growth he was gaining from Chip's lessons.

After lunch Marty went to the weekly engineering meeting in the conference room. With all of the engineers and Steve crammed around the inadequate conference table, he usually ended up jostling elbow to elbow between Kunal and Kyle, feeling like a two-hundred-twenty-pound giant next to his slimmer coworkers.

The meeting went fairly quickly and wrapped up when Alan finished explaining how they needed to improve documentation. As everyone started to pack up their things, Alan glanced around the room and asked, "Marty, can you hang for a second? Aubrey, Steve, you too please."

Chapter 4: Mocha in Mocha

Marty raised his eyebrows, but nodded. His coworker Aubrey had been filling in for him with a couple of clients during Marty's vacation, and since Alan had asked them both to stay, Marty figured that must be what it was about. What could have gone wrong in his absence, though? That familiar post-vacation dread reared its ugly head.

As soon as the room was clear, Alan got straight to the point: "We have a pretty ticked-off client."

"Oh, who?" Marty asked, nervous.

"CXL," explained Aubrey. "They called us saying they were having performance problems with a couple of their main hosted applications." Her short brown hair was cut to hang long over her gray-blue eyes, so she had to tilt her head up slightly to make eye contact.

This can't be good, Marty thought. *CXL can be a delicate client.*

"When I got to CXL, it was ugly," Aubrey said. "The problem was intermittent, so I had a bunch of people pulling me back and forth trying to show me what was wrong. Half the time things appeared to be working fine, but the other half I could tell something wasn't one hundred percent happy."

Marty took a deep breath as Aubrey continued. "I confirmed it wasn't on the application side and that the internet provider wasn't having issues, so I started digging on the client side. I tried to attach to the firewall using our standard IP/port and our username and password, but it didn't work.

I checked documents to look up if there had been a change, but it was listed the same as the default."

Marty felt his stomach drop. He knew exactly what she was going to say but didn't interrupt as she continued.

"I went to the closet and found the firewall, but—it wasn't our standard. I've never worked on one of these, I couldn't log in and didn't have a backup of it. I was stuck. All I could do was reboot and hope that fixed it. Well, it didn't come back up. It totally bricked. I had to run back to the office and grab one of our spares and scramble to get the client back online. It was a mess."

Aubrey shook her head as she continued. "If it had been a standard firewall, I could have simply restored the last backup to one of our spares and had it back in place in short order. But I had to fly blind and work through the problems to get them up and running again. It took way longer than it should have, and they weren't happy about it."

"That's an understatement," Steve added. "This is something that should have been done in a couple of hours, and we would have come off the heroes. Instead, they were impacted for *days*. Part of our standard is to have like models so we can swap! This made us look incompetent, and we can't charge any of this extra work!"

"It was pretty unprofessional," Alan agreed, crossing and then uncrossing his arms. "Marty, why was this client set up like this? Why are they not the standard, and why was the documentation process not followed in such a critical spot?"

"I am *so* sorry," Marty said. All eyes were on him, and he struggled not to squirm. "I had to use a different firewall there. I just forgot to—"

"*Had* to?" Alan interrupted.

"They were talking about adding more VPNs down the line, and that'll need greater horsepower. I knew this would serve them better as they grew. I do almost all of their work anyway, so it seemed like a no-brainer."

"You really can't 'cowboy' decisions like that," Alan said. "We have standards for a reason. We have spares so our clients don't have big negative impacts caused by technology failures. And the processes to document critical information are there to benefit the client *and us*. All of us."

"I'm really sorry it caused a problem, but it *is* a better setup for that client," Marty argued.

Steve snorted. "You're missing the bigger picture. The client wants a result; we want a result. If we fail to deliver that result, regardless of whether you think that component was 'better,' it's a failure. It's not better if it ends up with three voicemails from their president and several angry emails from their point of contact in my inbox. They lost a lot of productivity—it really cost them—and us."

Marty agreed. "I'm sorry for the problem it caused."

"Okay. Thanks, Marty. I guess that's it for now," Alan said.

As he walked out, Marty tried not to let any anger show on his face. But inside, he was frustrated, to say the least. *That was the best firewall for that client! Do they not want me to use the best tool for the job? I'm a tech—a damn good engineer. I used my expertise and evaluated the situation! I considered the standard, but this one is better.*

Marty headed to his office, his thoughts running. *Well, it was better for what the client was going to grow into. And as for the processes, if Aubrey had just checked in with me, I could have told her what was up. She could have called. It's not like I was on the moon.*

Marty had planned to meet Chip for a midafternoon coffee the next afternoon. Marty really needed this—Chip always seemed to see things from a perspective that Marty had not considered. Right now, Marty couldn't see what the other side of the CXL mess could possibly be, but he had a feeling Chip would surprise him.

When Marty walked into the upper-end coffee chain, he was greeted with soft, nondescript music and the rich smell of roasted coffee. The place had soft chairs, tables, and a few people wearing headphones while working on their laptops.

Apparently, the midafternoon pick-me-up crowd was in full swing. The majority of the customers were forming a sizable line, mostly made up of women in yoga pants and guys in shorts, waited on by a bustling staff with enough

Chapter 4: Mocha in Mocha

tattoos, piercings, and bright-colored hair that it could have been a body-art convention.

Instead of hopping right into the line, Marty scanned the place to see if Chip was already there. Sure enough, Marty found him in a corner, taking up a big squishy chair. Marty walked over and said hello as he dropped his backpack at the side of the armchair opposite Chip. Marty took off his cap, running a hand through his hair.

"I see you beat me here—so much for me buying you that cup of coffee," Marty teased.

In front of Chip, on the table, sat a big mug and a triangle-shaped pastry with pointy edges.

"I saw the line growing, so I decided to hop in and beat the rush," Chip explained.

"Ah," Marty said. "Well, thank you again for taking the time to help me. I'm sorry to waste some of it standing in line. I've been looking forward to talking. Let me hop in line and be right . . ." He trailed off, looking at the long line, and then grinned mischievously. "Actually, I've got an idea."

Marty sat down across from Chip and pulled out his phone. He swiped through his apps with a smirk.

Chip raised his eyebrows. "Seems like a strange time to play on your phone."

"I'm going to cheat the system. Since I don't get a whole lot of time with you, I don't want to waste it standing in line behind a crowd. I'll order it on the app, and they'll call my name when it's ready." Marty found his favorite drink—an iced mocha almond-milk latte—and put it in the cart. He made a flourish as he pressed the order button. "Aaaaand . . . done!" He put his phone down on the table, triumphant.

"Smart," Chip said, nodding his head in acknowledgment.

"Smarter, not harder," Marty agreed.

"Now that the caffeine is on the way, tell me, how goes the battle?" Chip raised his own cup and took a sip, settling in to listen.

Marty sighed and adjusted his posture in the chair. "In truth, it feels like one step forward, two back. Real success feels like it's always being yanked away. Yesterday was not one I want to repeat."

"That doesn't sound fun. Tell me why," Chip pressed gently.

"I have a client where I'm pretty much the engineer that does everything for them. A while back they needed a new firewall, and I guided them to the best solution to solve their need. Well, wouldn't you know it, while I'm on my vacation, they have a problem that ends up being related to that firewall."

Chip raised his eyebrows. "This is engineering, Marty. Everyone knows that problems happen. Especially when it's least convenient. So what was the issue with this one?"

Chapter 4: Mocha in Mocha

"The firewall wasn't the 'company standard,'" Marty said, making air quotes. "Since I was on vacation, they didn't know how to deal with it. It ended up being a big problem for TSC and the client. The worst part was that they could have reached me! They didn't even try. And I easily could have steered them to a faster reso—"

"Marty!" a barista with green-and-pink hair called out across the crowded room.

Marty instinctively popped up midsentence and walked over to the young woman. "That's me." He grabbed the drink, pulled a napkin from the dispenser, and wrapped it around the cup as he walked back to Chip.

Marty sat, took a sip of his drink, and made a face.

"What's wrong?"

"Ugh, I don't think this is right." Marty carefully swirled the drink, hoping that it was just a mixing problem. He took another sip and pulled a puckered, frustrated face. "Too much sweetener, not enough chocolate. You'd think with so many stores, and them charging this much, they would have an exact recipe that every store would use. Instead, I'm never sure of what I'm going to get. It's too sweet, it has too much ice, or it has no chocolate. How can you forget the mocha in a mocha? This really chaps my . . ."

Marty stopped when he saw the twinkle in Chip's eye. "Oh, don't smirk at me!" Marty scoffed. "I know that look. I'm talking about coffee, Chip. There's no way this relates to engineering."

Chip's eyebrows raised. "Oh?"

"Engineering is *completely* different," Marty objected.

"Is it now?" Chip did his best to hide a grin behind his coffee cup.

"Don't tell me you paid them to mess up my coffee for another of your 'learning moments'?" Marty shook his head.

"I didn't think of it until I heard you ranting about recipes, but now that you mention it . . . I should have. It's just too perfect." Chip's green eyes flashed in delight.

Marty went to take another sip of his coffee, then thought better of it and defiantly put his cup down on the coffee table.

"Sorry." Chip bellowed out his signature laugh before taking a drink of his own, presumably fine, coffee. "Standards are exactly how to build the consistency you want from your coffee."

"Okay," Marty conceded, "I get how coffee can be standardized. Hell, *that* should be easy. But engineering? No two customers are the same. No two engineers are the same. There are so many great solutions out there, and finding the best options for the client is a big part of my job as an engineer and consultant. My clients appreciate my tech expertise. And it's part of the job that I love doing!"

Marty nudged his disappointing coffee, a little despondent. "I don't know, Chip. It just . . . feels so wrong to restrict yourself in the name of uniformity. Do you really not want to use the best component for a client just to make it align with a standard?"

"Marty, I get it." Chip gestured to punctuate his point. "You're a smart guy, and it *is* engineering and problem solving. It feels good to be the author of something—to map it out, implement it, and see it work. There is ego tied to that, and that's not a bad thing, per se."

"It does feel good," Marty agreed.

"So let me ask you this. You love that feeling of helping the client. What does the client want in the end?"

"They want to engineer the best solution for their business," Marty answered with no hesitation.

Chip shook his head. "This is where I need you to set your ego aside, Marty, and think about the big picture. Remember, *their* interpretation of success matters. Don't insert yourself. What does *the client* really want, overall, for their business?"

This time Marty thought about it before answering. "They want what's best for their business. They want to be able to work without problems. They don't want to be down," Marty said, thinking of those angry emails from CXL.

"Ah, yes," Chip said. "To reliably transact business in the manner they have come to expect. Think, if there is any impact to their ability to work, how do they react?"

Marty chuckled. "They get upset fast. Really, really fast."

"There you go. Predictable and reliable—that's at the top of the list." Chip finished off his drink and set the cup down on the table. "Now let me ask you about your company. Does TSC want to install the latest systems, components, and software on important systems?"

Marty rolled his eyes. "Hell no."

"Why not?" Chip asked. "Those are new, exciting, and challenging things to work with, and they lead to new sales. Why wouldn't TSC always push that to their clients?"

"New things aren't proven," Marty explained, though he had the feeling Chip already knew the answer and was leading him to it. "There are a lot of unknowns."

"And if you implement those variables, what's the impact?" Chip asked.

Marty took a sip of his drink, forgetting it was too sweet, and made a face. "Well . . ." He paused, considering, then said, "It causes a lot of problems." He tilted the drink as an example.

Chip smiled at Marty's reaction. "So the company wants to limit problems too. To be as predictable and reliable as possible."

"Well, yeah." Marty sighed, seeing where Chip had led him. "And I guess that's fair."

"Hold on." Chip held up a finger. "Let's stay on the company for a little longer. Does the company care about efficiency?"

"Of course. They want us to do things as efficiently as possible."

"And how easy is it to be efficient if everything is new?" Chip asked.

Marty leaned in closer, fully engaged. "Well, you can't be."

"To become good at something, you have to practice it—you need repetitions, iterations," Chip said. "You have to lather, rinse, repeat until you work out all the kinks and know how to avoid—or minimize, at least—any problems. Companies and clients love best practices. Why? Because they want—"

"Predictable, reliable results," Marty said, cutting Chip off.

"You got it," Chip said, smiling. "Without standards and followed processes, best practices and efficiencies don't exist."

Chip took a bite of his pastry, savoring it for a moment, before he continued. "You know, standards and process work for you too. They help employees have a better quality of life. You have fewer solutions to deal with, so you can have a deeper knowledge with the technologies and become more proficient with them.

"With that tight set of standards," Chip continued, "more team members gain greater experience to help when you get into difficult situations. Plus, it makes for better vendor relationships, where there's even deeper support expertise to help. There are simply fewer unknowns and more people to help should there be a bump."

Marty nodded. "Other engineers from the company *should* have been able to take over when I went away without needing to call me on vacation."

"Exactly. How do employees feel when they're out on an island without a lot of help?" Chip paused, then added, "Versus how do they feel when there are people who have their back? It's a big difference. It really impacts culture."

Marty nodded, soaking it all in, head spinning a bit.

"You're in technology, Marty, and by default, there will be variables. The more variables, the more difficult the problem. If you can reduce them, it makes the job a little easier." Chip wiped his fingers on a napkin. "Standards and processes give you an advantage. They help you produce great results consistently, and then improve them."

"It's funny." Marty looked up, thoughtful. "I've always felt standards and processes were more for the IT engineers who didn't have the level of expertise to figure out a solution on their own. I never saw the benefit at my level." Marty blew air out between his lips. It was hard to admit that he might have been so wrong in this area, but Chip had a solid point.

Chip used the pause to finish his pastry before saying, "As for using your engineering skill to solve problems . . . let me add this. Your company needs people to step up and help with standards and processes. 'Standards and processes' doesn't mean innovation stops. You're not expected to take a passive role. The company needs people to set the standards and write and define the policies and procedures—to train others and improve them."

Chip leaned forward, passionate about what he was sharing. "It's on you to work to improve TSC's standards and processes. If you find a better way, apply that across the board. If you discover a new component, there is an appropriate way to evaluate it to see if it should not be the new standard." Chip grinned. "In fact, I can sum it up using your own words."

"My words?" Marty asked, surprised.

Chip nodded. "I believe you said 'smarter, not harder.' Using standards and processes is all about working smarter."

Is he right? Marty wondered. *Have I really not seen how this is better for me? I guess this does make sense at one level, but could I really improve by better leveraging standards?* It was a lot to think about. "I'm gonna have to chew on this one," Marty admitted. "But I see where you're coming from."

"You always need to remember you are one person in a larger team. To succeed, you need to reduce the variables and use systems that produce consistent, expected results. That's a key to build trust with your clients and peers and the company," Chip said.

"You show them that every time they order the drink,

they get exactly what they ordered." Marty tipped his drink in acknowledgment of the point, and Chip nodded.

"Are you going to order a different drink to replace that, um, one-of-a-kind creation?"

"No," Marty said. "I feel like drinking this has something to teach me." He took another sip and made a face. "Then again . . ."

Chip boomed out his laugh, drawing the attention of several of the baristas. "Oh, say. The University of Houston is your alma mater, isn't it?" He gestured at Marty's old white cap, which proudly displayed the UH logo despite being a little worse for wear.

"It sure is."

"Perfect! I've got a pair of tickets for the Rice game next month. My wife has book club and has no interest in missing that. How she chooses soup and gossip over a game beats me, but her loss. Do you want to join me?"

"I'll never turn down a chance to go to a Cougars game," Marty said, delighted.

"So, how 'consistent and predictable' do you think the Cougars will play this coming season? " Chip asked.

Marty laughed. "Okay, okay, I see your point!" He held up his hands as if to defend from an assault. "I'm listening. Tell me more about standards and processes."

Lesson 4: Standards and processes accelerate your success.

Chapter 5

THE MORE-EXPENSIVE OPTION

On a Thursday afternoon in late September, Marty left work to drive into downtown Houston. He headed to the offices of Julie Timms, a fairly long-tenured client. Julie worked for JMK Financial, located in a high-rise building in the Skyline district on Louisiana Street, which was dubbed "Mahogany Row." The buildings there were occupied by either financial or energy companies that seemed to drip success.

A month earlier, Julie had called Marty and asked him to create a proposal for some upgrades they needed and renew the managed services agreement. Around the same time, Marty had another one-on-one meeting with his boss, Alan. The managers had noticed the progress Marty had made on his performance improvement plan.

After his managers closed his PIP, Marty breathed much more easily. But he was keen to keep progressing. If he could win this bid with JMK, Marty would have another notch to show his managers that he could contribute at a higher level.

Marty had been excited about the proposal he had compiled for JMK. But just one week earlier, when he delivered the two-page proposal and diagram, Julie's response had been a little underwhelming.

As she took it, she looked across her desk at him and asked, "Is this it? It looks a little thin. Does this have everything you recommend for us? Are you sure you've considered our future needs?"

Marty was a little taken aback. He told her that it was short but comprehensive. The document included everything she needed to be able to see the value that TSC could provide. "Besides," he said, "you know how well we support you. You have no surprises with us; you know the level of service we provide. Our customer satisfaction ratings with your user surveys have been stellar. No bad surprises."

Julie thanked him and promised to be in touch.

Now, driving to the JMK Financial offices, Marty's chest filled with anticipation. Julie had called him yesterday, asking to meet in person.

Marty knew a few other companies had vied for the bid, but despite Julie's dispassionate response last week, he was confident none of the competitors would deliver with the same level of insight and awareness that he had. After all, TSC had been helping JMK for over five years. They had positive reviews on their services, and he was Julie's go-to

Chapter 5: The More-Expensive Option

guy. From what he could tell, they had all the advantages an incumbent could have.

Marty had the cost of the new infrastructure mapped out. It was a little steep, but he was ready with his reply: he had scoped it to be ready for them to scale. The amount for the project was just under $70,000, with two-thirds of that in gear and the rest in services. With this project, plus managed services at around $8,000 a month for three years, the value of the contract to TSC would be above $350,000. It would be great to bring that in to help make up for the loss he had caused on the Firefly onboarding.

Glad I won't make that mistake again, he thought. *With Chip's lessons, I'm growing into a stronger contributor for TSC.*

With extra pep in his step, Marty took the elevator up to the twenty-sixth floor and entered the opulence of JMK Financial. It was in stark contrast to the offices where he spent most of his time—everything was made of real mahogany, down to the skirting boards. Everything smelled like leather, wood, and fresh flowers.

The reception desk was a huge slab of granite that made the receptionist look small behind it. The view out the windows was of the vibrant downtown core, with Sam Houston Park in the back adding a splash of green. Almost every office had a stunning view, with the exception of a few admins just inside the large outer ring of offices.

An assistant met Marty at the reception desk and took

him to the offices. As they walked through, he caught a glimpse of the office kitchen's white granite counters—and a very large chrome espresso machine that rivaled the coffee shop he usually went to. *Leia would kill for a kitchen like that at home!*

Julie's office was slightly less ornate than those of the partners, which he glimpsed briefly as they walked past, but still stunning. The door, desk, and credenza were all made of the same dark, rich mahogany that gave the area its namesake. Julie had two monitors on her desk. A small, fresh flower arrangement sat on top of her large mahogany lateral file cabinet.

She stood and smiled warmly when Marty came in, but there was something hesitant in the way she invited him inside that gave him pause. "So glad you could make it," she said, surprisingly formal.

"Thanks for the invite," he said as the assistant left, closing the door behind him. *This is it,* he thought. *Just take a breath and try not to look too eager. With our history here, we're a lock. We are getting this . . .*

Julie gestured to the seat in front of her and sat down as Marty did. "Listen, Marty. I was really pulling for TSC. I know you guys do great work, and you have the technical skills we need," she glanced down at her desk. "But the management team decided to go with another company."

Marty tightened his grip on the arms on his chair, momentarily too stunned to speak. Finally, he found his

Chapter 5: The More-Expensive Option

voice. "Uh," he stammered. "Is there anything we can do? I know that we—"

"I'm sorry, Marty, but it's firm. The partners have made up their minds on this. I really shouldn't be showing you this," she said, sliding a large, bound presentation across to him. "But I wanted you to see what you were up against."

Marty picked up the hefty pile. It was a proposal, spiral bound with an ornate plastic cover. It had to be over twenty pages—maybe even thirty. "This is the proposal from the winning company?" he asked, stunned.

"That's right. The partners were really impressed with their understanding and awareness of what we need and our challenges as a financial company. They knew about our auditing needs—and about new compliance regulations coming down in the next year."

"$162,000?" Marty gasped. He had just reached the part of the proposal where they laid out the budget. He double-checked to make sure that wasn't the price for all three years, but no. They were charging *more than double* for the upgrade alone, a price he already thought was on the high end. And JMK Financial had *still* gone with them.

"You can take their proposal with you. Just please, do not show it to anyone. But I really think you should look over it—and understand what they're offering that just wasn't there in your bid. The strategy component, how they understood our business needs—not just the technical parts. The way they anticipated our future needs, laid out options as our business grows and changes . . . It was really compelling."

Marty, recognizing the kindness Julie was doing, tried his best to rally. "I really appreciate you giving me this," he told her. "I'll definitely . . . yeah, I'll read it over. I think there's a lot to unpack here."

"I'm sorry we won't be working together anymore," she told him. "But you never know, we might intersect again at some point."

"Yeah. Thanks." He stood up, and she mirrored him. She watched him as he let himself out.

More value at more than double the cost? he thought. *We're suggesting almost the same technology components. How is paying more . . . ? I don't get it.*

Marty was dying to talk to Chip, and luckily they were going to meet for the Cougars game that Saturday.

Marty picked Chip up at his house, and they rode in together. While he waited to talk to Chip about the JMK loss, Marty's mind still raced. *How did we lose that bid? What did I do wrong?* He didn't want to bring it up right away and ruin the game, though. He would wait until after.

Instead of talking about work on the ride in, they talked about the players on the team that year, who was likely to make it to the NFL, and whether they felt the coaching approach and strategy had been putting them in the best position to win.

UH was playing Rice, and the first half went by

Chapter 5: The More-Expensive Option

quickly—especially with beers and hotdogs in hand. But Marty was distracted, thinking about work, and only gave the game half his attention. It was a defensive battle with a lot of three-and-outs. At the half, UH led 6-3. Marty knew if he was going to enjoy the second half, he had to get his work problems off his chest.

During halftime, they decided to stretch their legs and return the beers they had "rented." Marty used the walk to the restrooms as an opportunity to inform Chip of JMK Financial's departure from TSC.

Chip asked questions as they returned to their seats for the start of the second half. "How much more were they charging?"

"Almost $100,000 for the upgrade alone!" Marty answered.

Chip's eyebrows went up, and he nodded thoughtfully. "And what did they say the partners valued?"

Marty answered, but the crowd's cheers were too loud. "I didn't catch that. The crowd was too loud—" Chip was cut off halfway through his sentence by another deafening roar.

They both laughed, and Chip shrugged. "Let's pick this up after the game, on the way home, when we don't have to shout."

In the third quarter, the pace picked up, with the teams producing three touchdowns and one field goal, with Rice taking a 17-16 lead.

"Both teams made adjustments on offense, and now the defenses are scrambling to keep up," Chip said.

Before they knew it, they had reached the two-minute warning in the fourth quarter. Rice led 31-30 and had the ball. It didn't look good for UH; neither defense could stop the other's offense, and Rice was killing the clock. But a holding penalty called back a would-be first down and put Rice in a down-and-long position they could not overcome. The Cougars defense finally held and forced a punt.

A chance for hope: UH had the ball with 1:45 on the clock and one timeout, and needed to gain about forty yards to get within field goal range. Rice's defense tightened, and UH had to fight for everything it got. If UH could get to Rice's forty-yard line, they had a chance of winning.

UH resorted to quick throws and sweeps to not only move forward, but also get out of bounds and stop the clock. Fans were on the edge of their seats. After ninety-five seconds and their last timeout, UH found itself at Rice's forty-two yard line. It was third-and-four with ten seconds on the clock. Time for one more play to try and gain yards to set up the kick.

They ran a quick pass to the left—the receiver managed to carry two defenders and make it to the thirty-eight yard line for a first down before he was tackled in bounds. The clock stopped to reset the first-down marker, but there were only four seconds left.

Chapter 5: The More-Expensive Option

A flurry of activity ensued: the referees reset the chains and set the ball in place, and UH's offense sprinted off the field while the kicking team ran on. The referee blew the whistle and started the game clock as both teams were getting set. UH had no timeouts, and the clock was ticking. Three . . . two . . . The holder looked at the kicker, who nodded he was ready. The holder turned to the center and pulsed his hands to call for the ball, but the whistle blew just before the ball was snapped. The kick went up and through, but it was too late. Time had expired.

The game was over. Rice, 31; University of Houston, 30.

The fans were restless and disappointed as the players left the field. Marty and Chip stayed in their seats, waiting for some of the wave of humanity to clear out before fighting their way back to Marty's truck.

"Well, that was a letdown," Marty said with a sigh.

Chip let out a big rush of breath. "You can say that again. Talk about not seeing the forest for the trees! I would hate to be that receiver. For a split second he must have been so happy he made the first down . . . then, *boom*, he realizes the situation he put them in. Seems like for that critical moment, his instincts to get the first down overrode the game situation they were in."

"Ugh," Marty agreed. "If he would have just gone out of bounds instead of fighting for the first down, the clock would have stopped, and they would have been able to kick the field goal. We could have won the game. Bigger picture."

Marty was silent for a moment. When he spoke again, his voice was quiet. "I think I can relate to that guy."

"Oh?" Chip asked, turning in his seat to face Marty.

"He did everything he could, to the best of his ability," Marty said. "But there was just . . . something more important that he didn't know. That's how I feel about JMK. They made it clear that the other company took their business strategy needs into account and I didn't. I don't know their business! I don't know their industry requirements! We do technology—not strategy. Why would they even think we should do that? I don't even see how it's related."

"What a company wants and even expects from their technology has shifted a bit." Chip looked out at the stands. "Technology can be applied at different levels, and each level contributes a different amount of value to a business."

"What?" Marty blurted. "It's all the same technology. Same Microsoft, AWS, whatever."

Chip nodded. "Lots of the components are the same, but the approach makes all the difference."

Marty cocked his head. "I don't know. That seems like a little BS to me."

Chip leaned in. "You need to be aware that when you engage in work, there are basically three levels of approach.

Chapter 5: The More-Expensive Option

There's the technical: doing the work. Then the operational or efficiency: how the job gets done. And last, strategic: why is the business doing the work, and what does the business look to gain from it?"

Marty shook his head. "I see that there could be different levels when you divide it that way, but I don't get how understanding every client's business could even be possible. We support dozens of different types of clients. I can't know all that!"

"I agree that it's a large amount to learn." Chip stretched his legs, gauging the crowd as he thought. "That's why a service provider often will focus on specific verticals so it can really understand them, gain efficiencies, and bring benefit to those specific markets. Look at GES. Virtually all of our business is in the energy sector, and we understand their business really well. It lets us dial into that industry's operational and strategic needs. To add more value than our competitors. To stand out."

"This is overwhelming." Marty looked up and let out a huge sigh. "I've worked so hard to incorporate things you've shared, but this seems like too much. I can't learn all that, and even if I do learn one business type, there are dozens more to learn."

"I feel you, Marty, I do." Chip placed a hand on his own chest. "But before you get too discouraged, let me share a couple of secrets that will make this manageable."

He held up a finger. "First, the basics about each business

model, such as how it operates, gains, or loses efficiencies, produces revenue, whatever, don't change much between industries. Technology is constantly evolving, but this is pretty static. You learn it once and pretty much have it down."

Chip and Marty squished way back in their seats and awkwardly turned their legs to the side to let a group of people pass the other way through the stands. Several had their faces and bodies completely painted in UH red and white, which caused Chip to shake his head in wonderment.

Marty was fixed on Chip, waiting for him to continue, but when he did not, Marty prompted him. "And?"

"And?" Chip repeated, only half listening, apparently still amazed that people would go to such extremes to celebrate a team.

Marty chortled. "You said there were two secrets. What's the second?"

"Oh yes, the second secret. Believe it or not, to gain the largest benefit, there's really only one type of business you need to learn about."

"What? Which one? I have a ton of clients. Do you mean my largest client?"

"Well, no." Chip scrunched his brow in thought for a second. "Actually, yes, it is your largest client—the client you spend the most time serving. Who is that, Marty?"

Marty thought out loud, listing several of the larger

clients he helped support, but Chip slowly shook his head at each one.

"Think. Which client pays you?" Chip prodded.

"They *all* pay, Chip." Marty looked quizzically at Chip.

Chip elaborated. "Who provides you with growth opportunities, gives you a paycheck, your benefits, hired you?"

"Oh, you mean TSC?"

"You got it," Chip chimed. "The very business you work for. That's the most important one to start with. Too many people focus on company clients and miss the fact that the first company they need to be loyal to, to help succeed, is their own company. Understanding the basics of how your company succeeds, its business models, operational efficiencies, and such, gives you the insight you need to contribute at a higher level."

The crowd was down to a more manageable level, and Chip indicated with his chin that they should start toward the exits. They got up and made their way out of the stadium, walking single file up the stairs with the remaining crowd.

Marty paused at the top of the stairs, letting a big swell of disappointed fans pass by. "Okay, so it's fairly static, and I need to focus on my company first. But when I hear technical, operational, and strategic, I'm not sure how those are applied. Can you give me an example I can use to better understand this? Because honestly, I'm just an engineer. Most of that sounds way beyond me."

"It might be beyond your training or awareness at this point," Chip agreed, "but it definitely isn't beyond your ability or potential. You just need to understand how each of those layers approaches a problem or opportunity, and what they're looking to accomplish.

"Let's consider the three approaches," he continued as they worked to cross the stream of people moving directly toward the spiral walkways, like salmon swimming upstream. "First, we can look at a project as the different activities or technical tasks that need to be accomplished. This is where most service people hang out, looking at and working on tasks. Just solving technical tasks until they reach the end."

"That's where I spend most of my time, in the valuable stuff, getting the work done," Marty quipped, and Chip knowingly smiled.

They reached another set of stairs, where it was far less crowded. Chip continued. "The second approach we can use in looking at that same project is operation and efficiency. How efficiently are we doing those tasks, at both a time and technical level? Are the resources assigned the best choices for the work and business? Is the client being managed through the process? Is the work getting done more efficiently, which raises the effective rate realized for the company? What can be improved,

Chapter 5: The More-Expensive Option

what was learned, what documentation was completed? Did we minimize the negative impact on operations? All that stuff."

"Okay—so maybe those don't seem so crazy, but *strategy*?" Marty squinched his face in skepticism.

"Well, for strategy we need to look at why the company is doing the project. What's the company's goal when you take on a project?"

"The client wants it done?" Marty smiled. He knew that wasn't the answer Chip was looking for.

Chip chuckled. "Sure, okay, we know the client wants it, but think about it from the company's perspective. The business has to have a strategic reason as to why it's willing to do a project. Remember, it can say no if it chooses to."

"Well, I guess it's to bring in money, revenue." Marty paused, furrowing his brow as something clicked, and quickly added, "And margin."

"You got it," Chip grinned proudly. "The business engages on these projects to bring in revenue—and as you said, most importantly, margin. The money that remains to help the business. I remember one of our first conversations, Marty, when we talked about the fact that you're in the game of business."

"And margin is how you score," Marty said. The two men walked through the parking lots to reach their section.

"Right," Chip nodded. "And it is fair to say margin is

strategic. So to recap, when you look at a project, it can be viewed as technical tasks and actions that need to be accomplished, or the management and efficiency with which the project is delivered, and at the top strategic level, the company engages on projects to produce margin. Whether that's thirty or fifty percent, or whatever your margin percentage is, that margin that needs to be produced to meet the business's needs."

"But how does this tie to my approach?" Marty asked. "Those levels are there whether I see them or not."

"Yup, they are," Chip nodded. "But you see, if you're thinking just of the technical, you tend to stop there, just working on technical. If you see the operational level, you aren't just doing the tasks, but also being mindful of efficiencies, performance, and everything else. And at the strategic, you're thinking at the business level.

"Think of it this way, Marty," Chip continued, as he walked slowly. "If you're in charge of a project, you have the responsibility to not only complete the project but also deliver the target level of margin or better back to the company. That way, you're thinking about what the business desires to accomplish and doing your best to make it so."

They walked in silence for a little while. As they approached where the truck had been parked, Marty pressed the remote button and the truck flashed and beeped to reveal where it was hidden, behind a larger vehicle.

As they opened the doors, Marty broke the silence. "And all of that is tied to my approach?"

"That's right. I'm going to say these three approaches again out loud, and I want you to pay attention to the shift in what you think about with each approach," Chip said. "Are you there to fix technical problems, are you looking to help the company operate more efficiently, or are you looking to help the company succeed in business?"

Marty thought this over as he put the key into the ignition and turned it. The truck vroomed to life. "Wow. Who knew that approach alone had so much impact?"

Over the next few weeks, Marty started to really apply this lesson to what Chip had already shared with him. He decided to stop thinking of himself as a technical problem solver and more as someone who solves business problems using technology. His vantage point was opening up, and he was seeing things at a new angle.

Lesson 5: Your value changes based on your approach.

Chapter 6

CATCH AND JUGGLE

Marty's phone rang on the desk beside him, and he resisted the urge to snarl at it. He had been in the zone, and the distraction completely broke his concentration. He shook his head, closed his eyes, and took a deep breath before answering.

Marty picked up the phone, careful to modulate his voice so none of the annoyance would sneak through. "Marty here."

"Marty—it's Jake at VSJ Inc." Jake was the CIO and the main point of contact for TSC's IT contract for the company.

"Jake! Great to hear from you," Marty said, though he was surprised. They had an established cadence for communication, and it wasn't like Jake to call outside of that—he was very structured.

"Do you have a couple minutes to talk?" Jake's voice sounded more somber than usual, and Marty immediately sat up straight. *Not today,* he thought.

Chapter 6: Catch and Juggle

"What's up?" he asked.

"I've just come out of the executive meeting, and I need to tell you we've decided to engage a different IT firm to support the executive team."

Marty took a deep breath as Jake continued. "I know we're still under contract with you for our main support, and that will remain the same for the rest of our support services at this time. But the execs feel they need a higher level of support than they're currently receiving."

"Jake, I'm so sorry to hear that." Marty grabbed a pen and anxiously started to flip it rapidly between his thumb and index finger. "I know it reflects on you and the decision to hire us, and I'd really like to see if we can fix things before you move to another vendor."

"I'm sorry, Marty. I wanted to argue your case, but honestly . . ." Jake paused, "I can see where they're coming from. The feeling was pretty much universal in the room. They don't feel you're responsive to them; they said the urgency is just not there. I had pulled a report to try and defend it, but in the report you can see that your team's response and resolution times *have* been dropping."

Marty winced. He rubbed his eyes and nodded before remembering that Jake couldn't see him.

Jake continued. "For the rest of the company, we can manage it, but the execs . . . I get that they can be a bit impatient, but they need a higher level of service than is being delivered. They just don't have the time to waste."

"But—" The word slipped from his mouth, then he stopped himself. Jake was right, and Marty needed to own up to that. "You're absolutely right, Jake. Our results have slipped, and that's on us to fix."

"I'm glad you see that." Jake's voice was still clipped, but relief bled through.

"Is there any chance they can hold off on the decision, give us a little time to fix this?" Marty asked. "I know what needs to be done, and I'm sure the team knows how to fix it. I can ask them to step it up for VSJ. We really value our relationship with you, and I know that we can do better."

"Marty, I can't fight them on the decision and hold any credibility. I know you have great intentions, but the results speak for themselves. I'm afraid at this time, that ship has sailed. But this is what I can, and will, do," he said.

Marty held his phone tight, eager for Jake's lifeline.

Jake's voice was calm and clear. "I'll make sure that our contract with the new company will only be a year, so it terminates at the same time that yours does. Six months prior to renewal of contract, we'll review your performance. If your results are where they need to be, we'll consider giving you the full contract."

Jake threw Marty another bonus, cordial but kind. "Plus, we would prefer to use one vendor, not two."

"Thank you, Jake!" Marty seized on the silver lining.

Chapter 6: Catch and Juggle

"That's absolutely more than fair, and I have full confidence that we can win back your trust."

"Okay, I hope that's true. I do like working with you." Finally, there was real warmth in his voice.

"I'll be in touch soon," Marty said before hanging up. He squeezed his pen tightly and leaned back far enough that his chair almost tilted over. *We've made them happy in the past, and I know we have the ability to achieve higher results,* he thought. *I know that we do. We just need to improve our processes. Six months . . . We have six months to get there.* It was like having another performance improvement plan—but for a client.

Marty's team could do it. He knew they could. And he had a feeling Chip would have some valuable insight into how to get there.

Marty checked his calendar. Perfect. He had plans to see Chip on Sunday at the New Year's event downtown. Their families would be there, but it would be easy to steal a few moments and pick Chip's brain. Chip would cut through the BS and get to the heart of it all.

Chip and Marty planned to celebrate the new year by heading to the festivities at Constellation Field. The event featured food, activities, and entertainment with a light display to enjoy at night, capped off with fireworks at midnight.

Armed with enough water, sunscreen, bug spray, snacks, hats, and sunglasses to start their own Walmart, Marty's family set off for the activities. When they arrived, Chip was already there with his wife, son, daughter-in-law, and grandson.

The teens—Linzcy, DJ, and Chuck—immediately ran off to hang without the old people, while the adults got introduced. Chip's son, Dallin, was a dead ringer for his dad. Chip's wife was a reserved woman, but she and Leia hit it off immediately, as they both shared a passion for home decorating. When Leia mentioned she was thinking about a kitchen renovation, the two women jumped deep into an animated conversation over the things they would do differently if they had the chance.

With the women focused elsewhere, the men agreed upon a patch of grass that was close enough to the stage where they could see the entertainment, but far enough away that they could comfortably talk and relax.

Since Marty was meeting Chip's son for the first time, the three of them chatted about typical stuff—how Marty and Chip had met, what they all did for a living, and how long they had been in Houston. It was standard Southern manners, and it kept them busy as Marty set up a blanket beside the Halls.

When Dallin excused himself to find the beer vendor, Chip asked, "How goes the daily grind?"

"Things have actually been pretty good lately, a lot better than before," Marty said, offering Chip one of the oatmeal chocolate chip cookies Leia had packed. "I've been applying a lot of the principles you've taught me, and they're helping, helping a lot. I think I'm actually getting the hang of them for the most part, but . . ."

"But?" Chip laughed his big, booming laugh. "There's a 'but'?"

"Oh, Chip, there's always a 'but.'" Marty grinned.

"Well," Chip said, "why don't you lay it out for me?"

"You don't mind? I know this isn't technically one of our coaching sessions." Marty was very aware of their families around them, but they all seemed to be entertaining themselves.

"Talking shop sounds great to me. Besides, I would rather have a root canal than watch a mime," Chip said through a cupped hand as he pointed to the stage.

Marty laughed. It was actually jugglers on the stage at the moment, not mimes, but he wasn't going to argue.

"I had a client call the other day and tell us that they're outsourcing part of their support to another company. It's just for the executives, but if we weren't under contract, I bet they would have moved on completely. The thing that's really frustrating for me is that they didn't even give us a chance. Instead of talking to us about our performance dipping, they went off and found a different company."

Marty shook his head, frustrated. "They've seen us do good work before. We're better than we've been lately. We know what to do; we've done it before. I wish they would have been a little more communicative and patient with us."

"Hmm," Chip mused. But before he could talk, the crowd roared. Chip and Marty both looked over to see what was going on.

On the stage, three jugglers were engaging the crowd. Two of them juggled knives, while the third, the barker of the group, had stopped juggling to work the crowd.

The barker played the crowd, which responded with fervor. "I can't hear you!" the entertainer revved them up again. "Do you want to see something really *dangerous and stupid*?"

The crowd cheered even louder.

"Okay!" he yelled. "Now that you're paying attention—let's get down to business by proving to you that this next part is both dangerous *and* stupid!"

Chip rolled his eyes and murmured to Marty, "Hold on for a minute. Let's talk after these guys hurt themselves. I'd bet the ambulance will be quieter than this crowd."

Marty laughed.

The barker held up a motorcycle helmet. Its windscreen was covered with black duct tape, making it impossible to see anything with it on. He moved toward one of the two jugglers as if to put it on his head.

The performance was choreographed well: the juggler did not stop catching and throwing the knives, but just moved his head to the side every time the barker tried to place the helmet on his head. Between throws of the knives,

Chapter 6: Catch and Juggle

the juggler shook his head no, pleaded, and feigned crying and praying, all milking the suspense.

"*Enough* of this," the barker yelled, turning to play the crowd. "What do you want to see?"

There were moderate cheers of "dangerous!" and "stupid!" from the crowd.

"I can't hear you!" he called, cupping a hand to his ear. "All together now! What do you want to see?"

"*Dangerous and stupid!*" the crowd yelled in unison.

"Folks!" he yelled. "I give you . . . *daaangerous and stuuupid!*"

The two jugglers squared to each other, and all three jugglers started chanting in rhythm with the crowd quickly joining them: "Yeah! . . . Yeah! . . . Yeah! . . ."

The "yeahs" coincided with when the juggler who was about to be blinded caught the knife: "Yeah!/[catch] . . . Yeah!/[catch] . . . Yeah!/[catch] . . ."

After a few knives to lock down the cadence, he gave a nod. In between the yeah/catch tempo, the barker put the helmet over the juggler's head. The juggler was now effectively blind.

"Yeah! . . . Yeah! . . . Yeah!" continued the chants, and the juggler caught and returned every knife in cadence. The crowd erupted in cheers, and the juggler took off the helmet and tossed it aside. All three took their bows.

The "yeahs" from the crowd turned into crazy applause and cheering. Even Marty was impressed.

"You don't see that every day," Chip remarked, politely applauding.

Marty clapped enthusiastically. "I have to admit he was right. That was pretty stupid." He grinned. "Amazing, though. Do you know how precise they both had to be to do that?" Marty raised his hand up, imitating catching the knife. "That knife had to be in that exact spot every time, or the performance would have ended very badly."

Chip's brow furrowed in thought. "What would it take for you to be willing to catch and juggle sharp knives that were tossed at you?"

"For starters? I would have to be able to catch and juggle!"

Chip waved the objection aside. "Say you're an excellent juggler, so your ability to juggle isn't in question. What would it take for you to be willing to trust someone else to toss knives to you while you were blindfolded?"

"I'd want to know they're really good at juggling!" Marty said.

"What's really good?" Chip asked, cracking open a can of soda. "What percent of throws would need to be spot on? Eighty? Ninety?"

Marty shuddered. "No way. For me to be willing to be on the receiving end, you'd have to put it there every time."

"So, understanding what needed to be done, or the

potential, or good intent . . . none of that is enough? They count for nothing?" Chip asked.

"For me? It would need to be there every time, or I wouldn't even be willing to consider it."

Chip gave Marty that knowing look that had become so familiar over their lessons for the last six months. "Hmm. That's pretty cold. You'd measure someone based on their results alone?"

"Okay, okay," Marty said, holding up his hands, "I think I see where you're going here . . ."

Chip nodded. "Yup, it all comes back to work. You can have all the potential in the world, all the raw ability, the understanding, but it's what you reliably produce that creates the real results, the proof in the pudding—the ol' cliché, if you will. And that can hurt because it can feel cold. But even though we know what we need to do, and have done it before, the client wants results every time—not potential."

"Because potential is not the same as results," Marty said, summing up Chip's point.

"You got it. And this applies to your relationships with your boss, your career, your company, your peers . . . almost everyone. In fact, come to think of it, Marty, your spouse is the only person who probably bought into you completely on potential."

Marty barked out a laugh. "Hey, no! That's too low." He smiled, nodding his chin at the stage, where the jugglers were cleaning up. "Do you just come up with these metaphors as you go, or do you sit around waiting for one that fits what your lesson is?"

Chip chuckled. "Oh, I have my standbys should the environment fail me. But I could go on. Customer service is like juggling; working with technology is like juggling; parenting . . ."

"I'd take you to the sewage treatment plant just to challenge you, but then again—I might never erase from my brain what you'd come up with," Marty said, laughing. Chip grinned and shook his head.

Chip's wife interrupted to say they were going to walk over to see the lights turn on in the stadium, and they rejoined their respective families before making their way in. Marty spent some time reflecting on what a year it had been—but when he found himself starting to focus on work, he told himself it was time to relax. For the next few hours, at least.

Later that night, when he and Leia snuggled up as the dark sky filled with vivid bursts of lights, he felt confident about work and ready to put these new lessons into practice to anchor his career growth.

Marty knew that next week, next month, next quarter—they would all bring new challenges. The only way to find out if he was ready to meet them was to try—and see what happened.

The next week, Marty was working on a project and

Chapter 6: Catch and Juggle

ran into a stumbling block: he didn't have all the resources he needed to finish the project on time. He knew Kunal had the perfect skills for the role, but when he checked in, Kunal told him that he was busy building workstations for another client. It seemed like such a waste. Anyone could do the work Kunal was assigned to, but only Kunal could get Marty over this resource hump.

He went and knocked on Alan's door.

"Come in!" his manager cheerfully called out.

Marty popped his head in the doorway. "Hey, Alan, I've got a problem on my project that Kunal would be perfect for. Is there any chance we can move him off workstations for a bit?"

Alan opened a few documents on his computer, checking on timelines, and shook his head. "Sorry, no, I need him where he is."

"Damn, okay. Thanks." As Marty turned to leave, a thought came over him and he paused. "Out of curiosity, how many workstations do we do per year?" he asked.

"I'm not certain, but I would bet about six hundred or so," Alan guesstimated.

"And to confirm, a build takes two to four hours each?" Alan nodded.

Marty did some quick math in his head. "So with an average of three hours per build, that would be about eighteen hundreds hours a year on workstation builds."

"Sounds about right," Alan said.

"Okay. Thanks, Alan." Marty left, the numbers spinning through his head. He needed Kunal on this project. But with eighteen hundred hours in workstation builds, the company practically had a whole person tied up in that every year. There had to be a better way. And Marty had an idea.

A few days later, Marty knocked on Steve's door. "Hey, Steve, can I run something by you for a minute?" he asked.

Steve was working at his computer, but he turned away from the screen. "Sure, Marty. What's up?"

Marty closed the door and took a seat at the chair across from Steve. He was a little nervous. He took the chair that, just a few months ago, he had sat in when he met with Alan and Steve to discuss whether he had a future with the company. Today he was ready to not only prove his worth, but also show just how far he had come.

"It's about resource management. We build about six hundred workstations annually, at an average of three engineering hours per station. I've thought about it, and I believe we can be more efficient with those eighteen hundred hours—and produce an additional $100,000 to $140,000 in revenue in the process."

Steve leaned forward, cautiously impressed. "And how do you see this being possible?"

"We cut our hands-on time to deliver a workstation to about forty-five minutes by leveraging an imaging solution that can take a large portion of the work,"

Marty proudly answered. "About fifteen minutes of time would be needed to set up the workstation to receive the image and to package it up afterward. While the image is downloading, the engineer is freed up to do other work. At the end we would need another thirty minutes or so to check off the details to make sure the client is happy. We would spend about four hundred and fifty hours of engineering to produce the six hundred workstations."

"Okay, but how do we make any money if we cut our hours down that much?" Steve asked. "We charge for that time."

But Marty was ready for the question. "The increase comes this way: the client is currently being charged a random two to four hours, costing $350 to $700 per workstation, with an average of three hours and $475. We could charge a fixed fee instead. It would be lower than the average, let's say $400, so the client wins as well. The exact figure is something you would set with the client, Steve, as that's not my area.

"With four hundred fifty of those hours," Marty continued, "we realize $240,000. The remaining thirteen hundred fifty hours saved could be billed at the standard $175 an hour, which would come to $236,000. Currently, the eighteen hundred hours are generating $315,000. With this strategy, those same hours would generate $476,000, an increase of about fifty percent."

Marty was on a roll, passion threading through his voice. "There is a caveat. To do this, we would need some additional software. I looked into a toolset that can do this type of work, and it has a free eval that I could try. I've talked to Kunal, and he's willing to work with me on this—he would love to get out of the number of workstations he has to do. If the tool works, it will cost about $20,000 a year, so that would cut into our savings, but still leave us making an additional $141,000." Marty leaned back, eyes shining, thrilled with his own pitch. "What do you think?"

Steve just sat there. He didn't say a word.

Marty felt a thread of panic rise through his guts. *Oh crap. Did I overstep? Did I not explain it clearly enough?*

Steve shook his head. "Wow. I have *never* had an engineer come to me with anything like this. Increased revenue and close to $150,000 more margin just by doing this?"

Marty nodded. "I believe so."

"Great Scott, Marty!" Steve's face broke into a huge smile. "This would be amazing—you need to make this happen. This would be good for all of us." He looked Marty in the eye. "And I mean *all* of us."

Marty felt all of the tension of the past year flow away. He had done it.

For all of them.

Lesson 6: You are measured based on your results.

Chapter 7

SAILING

The morning sunlight poured through the window of Marty's new office. It wasn't as large as Steve's, but there was enough room for his desk and chair and a visitor's chair. It had a big whiteboard where he could brainstorm new initiatives. And it had a view. Granted, it was of the parking lot, but it was a window. On clear mornings, the powder-blue Texas sky seemed to stretch into infinity.

Marty had just taken a call from his wife trying to pin down the family's Easter dinner plans. With their kitchen in shambles, they would spend it at her sister's. *Oh well,* he thought. *I guess I can't win them all.*

He couldn't believe it was spring again. So much had changed. A year ago, he was so frustrated that things weren't where he wanted them to be. He had been a bit jaded, truthfully, with a career where success felt unattainable.

Unfocused, Marty stared out the window as his mind raced over all the things that had happened in the past year or so: the performance improvement plan, learning the things that really hurt the company and how to avoid them, the actions that could build and destroy trust and professional competence, and learning the key business drivers for a professional services company.

He shook his head as he thought about how relatively small some of the changes were that had made such a big difference for TSC and its clients. Small changes had produced big results, and truth be told, he wasn't left out of the benefits, either. The improvements he made had led to his promotion. Three months ago, Steve promoted Marty to engineering manager with his own team. Leia had celebrated her husband's promotion by dusting off the kitchen renovation plans with an all-too-mischievous grin.

On his way to the conference room, Steve stopped by Marty's office, knocking on the open door to get his attention. "Meeting is about to start. Ya' ready?"

Startled out of his reminiscing, Marty looked up. "Yes, yes I am." Today was the second-quarter management meeting.

Steve disappeared toward the conference room.

Marty gathered his laptop and pen. When he reached for the pad of paper, a bright white marble chip on a desk plaque caught his eye. *Chip!* He thought, and his throat choked up. *I can't believe how fortunate I am that our paths crossed. I don't even want to think about what it would be like if we had never met.*

Chapter 7: Sailing

Marty had been so busy with work that he hadn't spoken to Chip in three months. They had last spoken when Marty called to thank Chip and let him know about the promotion. With all of the work changes, school, and kids' activities—not to mention a nasty bout with the flu—they hadn't connected.

Come to think of it, I haven't seen him at any of DJ's soccer games, Marty realized. *I need to call him and take him out for a Jameson to thank him for all of his lessons. Hell, I need to send a case of Jameson. His help is really paying off.*

Marty left his office and headed into the conference room. As he sat down, he smiled about having more room without the rest of his engineering team jostling him at shoulder and elbow. With only six of them in the management meeting, everyone spread out their computers and papers around the wide table.

Everyone was talking, ribbing each other, when Steve said, "Okay, it's nine. Let's start."

Steve started in. "Normally we would just be reviewing how we did in Q1 and then making adjustments to Q2 to align with our annual plan. But there's an opportunity I want to try and capture.

"When Marty was promoted," Steve gave a nod to Marty, "both he and Alan completely freed me up to focus on growing the company, which has opened more opportunities than I had even imagined." He paused and looked around to meet each person's eyes before he continued. "We are at an inflection point, and TSC has an opportunity to do something bigger.

"After the last couple months of focused effort, we are about to land a couple of sizable accounts," Steve continued. "We've also connected deeper with several of our current clients. Great opportunities to provide more benefit to our client base abound—so much so that we need a major pivot from our current trajectory."

Marty felt like his head was spinning—but in a good way. It was cool to feel like he was a part of the company at a higher level. He wasn't just a tech nerd anymore. He was actually in the game.

As they reviewed Steve's changes for TSC's plans, the reality of how ambitious the growth plan was started to sink in. Marty's euphoria quickly faded. This would take a lot of work—more work than the six of them at the management table could orchestrate. To even get close to this vision, it was going to take everyone they had, and then some. The status quo was not going to cut it. Not even close.

Marty couldn't help but notice that the areas Steve was targeting for expanded services were directly in Marty's bucket of responsibility. It was clear that a lot was going to be solely on him to make happen. Steve confirmed it when he stopped and looked directly at Marty and said, "Marty, much of this is going to be on you to deliver."

Chapter 7: Sailing

Marty hesitated. He didn't want to have a negative attitude, especially when he was so new to the management team, but he had no idea how his team could hit these aggressive targets.

Steve finished, "I know we just completed our annual plan last quarter, but with this change of opportunity, we are going to need to adjust to hit these targets. Each of you need to pin down your plans and present them to me by next week. We'll then present the entire plan in two weeks at the company meeting."

Everyone stood up. The mood was more somber as they exited the room, and Marty wondered if everyone else felt as unsettled as he did about getting their teams lined up with the greater company targets. He held back in the room until it was just he and Steve.

"I'm not sure I have enough people who work at the level this requires," Marty said.

"You've made phenomenal strides in the past few months," Steve said, waving a hand to dismiss Marty's concerns. "I have faith in you. I know you're going to be able to figure it out and get it done for us."

"I appreciate your confidence in me," Marty said, straining to keep his voice even and strong as Steve nodded and left the conference room. But inside, Marty was reeling.

This was an incredible opportunity, but was he up to the challenge? *What if I don't have the ability to do all this?* Marty thought. *I'm going to screw up. This is going to be a* disaster. *There's no way we can get all of this done. The company needs more leaders and fewer followers. How the hell am I going to be able to pull this off?*

Then Marty smiled. What was he stressing about? Every time he'd hit a speed bump over the last year, Chip had been able to help him sort it out. All Marty needed to do was call Chip and this crisis would be averted.

After the meeting Marty went back to his office and started dialing Chip. Then, thinking that he didn't want to interrupt if Chip was at work, he sent a text message instead. He didn't hear back.

The next morning, a Saturday, Marty tried Chip again, this time by calling. He got one ring and a weird click or two before it started ringing again. But instead of reaching Chip and finding a solution to his problems, Marty heard a bizarre "pa ra ka low?" It sounded like a woman's voice.

"Hello? Hello—Chip?"

There was an answer, but it wasn't in English. Confused, Marty hung up and tried again, but the same woman answered: "Ne. Pa ra ka low." Marty hung up again. *This is strange.*

Sitting at the kitchen table next to him, Leia sipped a cup of coffee, and he told her about the strange call.

"What language was it?" she asked.

"I don't know—it all sounds Greek to me," Marty answered in a smart-aleck tone. "I need to talk over a problem with Chip. I'm going to rush over to his house now, see if I can catch him before he heads out for the day."

Marty left straightaway and drove to Chip's house. He had been there a couple of times to pick him up, though he had never actually been inside.

The house looked as he remembered it. The garage door was open, and he felt relief when he saw someone inside the garage that resembled Chip. After pulling up at the curb, Marty realized it was Chip's son, Dallin.

Marty got out of the car and walked up the driveway. Dallin was too busy taping up cardboard boxes to notice he had a visitor, and he jumped when he saw Marty.

"Oh! Marty, right?"

"Yes. Nice to see you again, Dallin," Marty said. "Helping your dad with some reorganizing?"

Dallin gave Marty an odd look, like he was smoking something. "You mean packing up the last of his stuff?"

"Oh, God. He's not . . . He *died*?" Marty put a hand on his chest, horrified. *Why didn't I check in? I can't believe I waited so—*

Dallin laughed, the sound reminiscent of Chip's big booming laugh, though not quite as earsplitting. "Are you kidding? He's healthy as an ox. He and Mom are in Greece. They're on the water somewhere, sailing around the Mediterranean."

Marty knew he must look silly, but he couldn't stop his

jaw from literally hanging open. "I thought this was his house."

"Oh, no, this is ours. They sold theirs last year. They were staying here until he finished up at his work." Dallin smiled sympathetically. "I take it by your expression that he forgot to tell you. They were so busy and frantic when they left. Way too many things to do."

"He . . . no, no, I had no idea," Marty stammered. "I tried to call him, but a woman answered." Suddenly the response he'd gotten earlier when he called Chip made total sense. It hadn't just sounded like Greek. It actually *was* Greek.

Dallin nodded. "Well, three months back, Dad's responsibility at his company finished up, and he was free. The plan had always been that when he retired, they were going to move to Greece for a few years."

Dallin smiled as he continued. "My parents had honeymooned there, and, well, it's their happy place. They thought it would take a few months to find a home in Greece, but it all just fell into place. They found a house right away, but they had to jump to get out there to close on it. They didn't even have enough time to pack up all their stuff." He motioned to the stack of boxes.

"Mom wanted a new kitchen," Dallin looked toward the garage entrance, "so they decided to hop on the boat and sail until the work is done. With spotty cell service on the water, Dad forwarded his phone to their new house.

You must have reached the housekeeper they hired to watch the place while they are out sailing."

"That's . . . great," Marty said, even though he was thinking, *No! How can he just disappear? I can't believe he's gone and I can't ask him anything. Hell, I can't even properly say goodbye.* Marty sighed, but said, "I'm happy for him."

"You can grab his address if you want. It's right here." Dallin laughed and gestured to a box as if he were Vanna White.

Numb, Marty's took out his cell phone and snapped a photo of the address label on one of the boxes Dallin was taping up. Marty couldn't believe it. Chip was gone, and Marty was completely on his own.

He thanked Dallin again and walked back to his car, but he wasn't really seeing or hearing anything around him. *What the hell am I going to do? How am I going to fix this without Chip's advice? Steve is counting on me. TSC is counting on me. I need to give a presentation in a couple days, and I have no idea how to solve this or what I'm going to tell them. What am I going to tell the team?*

Marty stared at his phone. He texted Leia and warned her that he would being going to the office today and did not expect to be home for dinner. Chip was gone . . . and Marty had to figure out how to be his own Chip.

Can I do that? Marty wondered. *If I can't . . . there's no one else now.*

It had been a long time since Marty had been in the office on a Saturday. The place was still and quiet. He didn't bother turning on the main lights. Instead, he went straight to the kitchen in the back and fired up the coffee machine—he knew he was going to need the extra caffeine.

With the coffee brewing, Marty opened the office fridge looking for creamer. He found it behind what looked like someone's leftover lunch that had been there for longer than he cared to think about. *Wow, does this need to be cleaned out,* he thought. But he knew that would just be a distraction.

He took his coffee back to his office. He flipped on the lights, blinking as the room came to life around him. The desk he was so proud to have earned. The whiteboard, already traced with tasks, projects, and a "love you, Dad" message in the corner that Linzey had left from the time she visited the office. He put his backpack next to the desk, plopped in the chair, turned to face the whiteboard, and grabbed a marker.

Somewhere within him, he knew there was an answer that just had to be unlocked.

He started to write on the board, shook his head, then erased it. Started again, but erased that. This cycle continued for almost an hour, until Marty just stared blankly at the whiteboard, chin in hand, getting nowhere. The answer must be within him, but what was it? He had never felt so incredibly alone. Ironic, considering he was now in charge of a large team of engineers.

Marty stood up, shaking his body to try and bring some energy back into it. He walked over to his desk and rooted around in his bag for a Diet Coke, choosing habit over coffee. As he put the backpack down, he noticed the org chart he had pinned up on his board. When he'd first got it, he focused on his position—the title above his name, the box around it all. Now all he could see were the names of the people whom he was responsible for, whose careers were in his hands.

They had families. Careers. Dreams. They were relying on him to help them find the path forward, to show them the way. They were all in this together.

He staggered backward, physically sinking into his chair with the power of the realization he had just had. *They were all in this together.*

Damn, he thought. *I'm still stuck in the attitude Chip was trying to shake out of me. I've learned the tools and business knowledge he was trying to teach me. But I was too focused on myself and missed this important lesson: Every player on the field needs to know the rules.*

Marty glanced at the ceiling, immersed in his thoughts. *I was drifting before Chip taught me what I needed to know. It wasn't that I wasn't capable. It wasn't that I didn't have the intelligence, the drive, or the ability. I just didn't have the perspective and understanding I needed. There is amazing opportunity for everybody in the company—I just need to help them be ready for and realize those opportunities.*

Marty fired up his computer, mind whirring. He couldn't wait to pitch Steve his idea.

In Steve's office, Marty stood as he presented, too full of energy to sit.

"There's a huge problem in our industry with the engineers and service people," Marty said. "There's a false sense of division when it comes to responsibility in professional services companies. Most service people believe it's their job to have the skill and deliver great customer service. Meanwhile, they think that it's management's role to make sure the company is successful and profitable."

Marty clicked forward to the next slide. "The problem with this belief is that every person who's involved in delivering service has a tremendous impact on the results. Obviously, that's true at a service level, but here is the big, missed point: Their daily choices and actions are what drive the *business performance level* as well."

"Our company is the sum of all of these individual decisions," Marty continued. "We need to make sure everyone making those decisions has the information to understand their impact—and make better choices."

As Marty shared the final slide, he resisted the urge to stick his hands in his pockets, instead taking a strong stance. "Everyone here has good intentions. Great intentions, in fact. We just need to give them some of the missing understanding, so everyone on the field knows the rules. If they can tighten it up like I did, they can do what the company needs from them—and more.

"Steve, I would bet you have at least two or three more managers and leaders in there, and with your growth plan we're going to need them."

Marty took a deep breath and stopped. He waited, eager and nervous, to hear what Steve would think.

Steve considered, then nodded. "Marty, this is a powerful approach, and I know you've just gone through this. But I'd be lying if I didn't say that I'm concerned that not everyone will buy into it all and make the change. I hope they believe you."

Marty nodded. "I understand that concern, and I believe the people will divide into two—well, potentially three—groups. The first, which is the vast majority, will learn enough to be able to reduce the areas that are big business penalties—the areas where we really hurt ourselves. This improvement alone is a huge gain; eliminating those performance killers will greatly increase company performance.

"The second group," he continued, "will be much smaller and will really learn and embrace the techniques to add greater operational efficiency and strategic business value. That group will be your new leaders, those who will really benefit the company and expand their careers significantly."

"Hmm." Steve was deep in thought for a moment, then prompted, "And the third group?"

"Well, there may be those who just keep making those painful mistakes and might not be long-term players on the team. It's really hard to win when people keep committing game-changing penalties." Marty shrugged.

"Well, Marty, it's clear you've thought through this, and I think you're right," Steve said. "It's going to be on you to show them that the opportunity for growth is real—and how to take advantage of it."

Marty grinned. "Don't worry about that. I've got this."

And he knew that he did.

Lesson 7: Amazing opportunity is in your grasp.

Epilogue

The sun was shining, and a light breeze blew for the final game of the season.

Marty sat in the high school stands with his wife, watching DJ play for the freshman team. Three years had passed since Chip left for Greece, and Marty had chosen to step up rather than down.

The game was close—one-to-one—and the second half was almost over. When a play near the side went out of bounds and the players shifted for a throw-in, Marty's phone rang, causing several close parents to turn and give him the evil eye.

"I've got to take this. I'll be quick," Marty promised his wife. He kept an eye on the field as he moved a bit away so as not to bother the other parents.

"Marty? It's Selena!" Selena was one of his mentees from work. He had been coaching her for the past several months and was so proud of her progress. Now her voice was fluting and excited. "I had to call and say it again. Thank you for the promotion and your belief in me."

"Well, you made it easy. I knew you could do it. I could tell how you embraced the concepts from the start," Marty said, delighted.

"It was all thanks to you," she said.

"No, it wasn't," Marty gently chided. "You were already a great engineer. You added the missing pieces to unlock your potential, and you need to take credit for what you did. Not everyone embraces it, but you did. It's all you, Selena. You deserve it."

"This takes so much weight off, Marty," Selena said. "My husband and I were wrestling with some financial constraints, and, well, this made it all so much easier. And I can't even tell you how proud my parents are!"

"Selena, I'm so happy for you. Let's celebrate on Monday, lunch. My treat. Is barbecue okay?"

"That sounds perfect."

"Great! Okay, I have to run now, I'm at my son's game."

"Oh, right!" she said. "Sorry to interrupt. Wish him luck! And thanks again, Marty."

Marty hung up with a big smile and took his seat. "Selena was just calling to say thank you for the new engineering manager position," he murmured to Leia.

"That's awesome, hon." Leia squeezed his knee, and he smiled. It was pretty incredible. He had been working with Selena ever since Steve had formalized the mentorship program that Marty had started. She had really absorbed it all quickly.

Epilogue

It reflected the general improvement in the company. In three years, TSC had more than doubled its revenue, and hiring was almost as busy. The company had a great reputation, not just as a great engineering firm, but also as people who could be trusted and who understood the needs of their clients and their work—and their results backed that up.

That improvement reflected in Marty's life too. He was no longer stressed about sending both DJ and Linzey to college. Leia had the house of her dreams, and on sunny mornings, Marty often watched the sun rise from his patio with a warm, flat Diet Coke.

It's funny, Marty thought as he watched the kids race across the pitch. *As humans, one of our core drivers is the need to feel like we matter, yet we're so often unable to see that we really do.*

Marty smiled as he settled back to watch the game. That was now his mission as VP of services: to make sure that every person on his team knew the basic rules for professional service business success, could see the opportunities before them, and knew that there was no place to hide in the service chain. Outcome, results, and career success were up to each individual and available to those who embraced it.

Every day brought new challenges—and new adventures. And he was ready to face them all.

Lesson Summaries

Lesson 1: You're in the game of business.

You may think you're in the tech support game or even customer service, but ultimately you're in the game of business.

How you deliver service and use your time at work has a major impact on the business's results.

The more you understand the rules and drivers for service business success, the more opportunity you will have to align your efforts to help the business at a greater level. If you align your work with those good business practices, you help your company win, and your value really grows.

If your actions work against good business practices, you reduce your value to your company, peers, the clients you work for, and most importantly your career success—even if you are working hard!

Lesson 2: Their interpretation of success matters, not yours.

Your interpretation and biases can lead you astray. You need to know what "they" view as success. It is up to you to confirm what their interpretation of success is. Remember that "they" means anyone you are serving and doing work for: your manager, company, peers, and client.

Stay on track with these simple rules:

- Learn what success really is. Do not assume you know. Confirm it. If "success" has not been made completely clear, share your interpretation in detail and verify that the definition would work for the individuals you are working for. By taking the lead and providing your interpretation, you create a starting point that can be adjusted by the person you are doing work for. Set the target clearly so there is no confusion.

- Form matters. It is not just "work effort," but effort to do it the way the company has asked. That means the details. And once again, "they" set the form/success criteria.

- Consistency counts. Just like in exercise, you get greater benefit from being consistent. You cannot binge success. It is really hard for others to trust what you are doing if you are sporadic.

- Effort is all from you. You cannot blame others or make excuses. You are the only one who can put in the effort for your results, and you cannot wait for the benefits and *then* put in the effort. Your effort comes first, before the results.

Lesson 3: Your competence is measured beyond your tech skills.

Those you work with often cannot measure or understand your technical skills or proficiencies, but they can and do assess the surrounding professional skills that are the hallmarks of professional competence. These actions quickly reveal if you truly are a competent professional and can be trusted or not.

Hallmarks manifest that you are responsible, knowledgeable, experienced, proactive, respectful, and reliable, and that you demonstrate a positive attitude and great communication skills.

Additionally, trust grows when people know that you are looking out for *their* success. Helping them succeed matters—not just doing a technical task. People like to surround themselves with people they can trust and push those away who create risk and can make them look bad.

Working to improve your abilities in the key attributes of a professional will almost always have a greater impact on your career than learning a new technical skill.

Remember: it is not just with a "client" where this applies. Peers, your manager, and your company also measure competence beyond your tech skills. Additionally, this has implications with your personal relationships.

You cannot take some tasks or days off, as reliability is

critical. Professional attributes are necessary in every interaction, from simple to the most complicated tasks. You can be the greatest technical mind, but when you fail to act like a true professional, your professional competence is called into question, and it is hard to trust you.

Lesson 4: Standards and processes accelerate your success.

Standards refer to the parts, components, and software that are used. As technology is so vast, an individual or company cannot know it all. Becoming more effective with a smaller set creates an advantage.

Process is how you get results in the most predictable manner. Process sets a minimum bar line of performance, one you can stand on and improve over time. Whenever you deviate from a standard or process, you put the result at risk by adding more variables.

Clients seek out service companies that can deliver predictable, reliable results and view documented standards and processes as signs of a mature service company.

As a company expands, and as more people are involved to produce the desired "result," more standards and processes become essential. Their absence stops companies from growing.

Standards and processes are not perfect, and they need upkeep and improvement. Companies need and appreciate those who create and improve processes because they help many people succeed and produce the desired result. If you contribute to helping others succeed, it opens up opportunities for you.

Standards and processes can feel like a blow to the ego, as there is a strong desire to "create and engineer" new solutions. To fully embrace standards and processes, you need to come to terms with what the client and company really value: reliable, predictable results. This does not come from "winging it."

Both your clients and your company want predictable outcomes. They appreciate this so much that clients will search for a firm that can deliver it, and companies seek employees who can deliver the same. Predictable, reliable results win every time.

Lesson 5: Your value changes based on your approach.

The larger the problem you solve, the greater the value you create. All companies have problems at the technical, operational, and business/strategic levels.

Your approach is defined by the understanding and intent you use to solve a problem.

To diagnose what your approach is, read these statements and note which one you relate to most:

- I solve technical problems.

- I solve operational and efficiency problems utilizing technology.
- I solve business problems utilizing technology.

Each of these has a required level of skills and knowledge to be effective. For example, to resolve a technical routing problem, you need to have solid understanding of TCP/IP, routing, the gear you are working with, and related troubleshooting techniques. If you do not have these, your ability to successfully resolve the problem is not high; you may not even recognize that it is a routing problem and consequently waste time in other areas.

The same logic applies to solving operational, efficiency and/or business problems. You need to have strong knowledge of related troubleshooting techniques to properly address them. If you do not have enough understanding, you will not be aware that the problem or need exists. So often there is great benefit within easy reach that is overlooked because we are blind to it.

The largest opportunity for you is to focus first on learning the basics of good operations and core business concepts for IT professional services and apply them to the company you work for. This helps you start seeing opportunities and increase confidence to start addressing them. By helping in these areas, you will quickly stand out, and your value to the company grows quickly.

You will need to find information and training that

gives you a good core understanding of these areas in an IT professional services company.

Note: If you are content with your current level in your career, approach, and compensation, there is nothing wrong in wanting to stay at that level. Competent professionals are needed at each level. When a person wants to be compensated for a higher level than they contribute or when a company puts someone in at a level that is beyond their current skill set, problems arise.

Lesson 6: You are measured based on your results.

You are what your results say you are. You can have all the potential in the world, all the reasons and excuses, but what you put out is what you put out. This can feel callous, but it is the truth.

Without results, you are selling faith and speculation, which are not as valuable as results. We often overestimate our contribution and value, but our production reveals the truth. You need to produce. People buy into results. Consider the value of these four levels:

- You understand what needs to be done and have a basic idea of what needs to be accomplished.

- You know the steps and have the skills to get it done and occasionally do it.

- You often are able to produce the result, but there are times you do not.

- You produce that result every time like clockwork.

Each level carries a different value, and expecting to get compensated or given a different level than you are producing is not realistic. Your value and the level of trust others have in you are based on your results.

Lesson 7: Amazing opportunity is in your grasp.

Small-business IT services companies need people who do not commit performance-killing mistakes, and they search out those who are able and willing to take on and effectively fulfill leadership positions.

The growth of most small service companies stalls because there is a dilution of solid leadership. Leadership opportunities are not limited only to the technical level, but also are needed to help with operational and strategic problems.

Those who are willing to learn the core elements of solid business practices can stand out and make a huge impact. When you can deliver results, the roles you are presented with grow exponentially, and your income will keep pace. Leaders and owners will recognize the benefits you are providing, and your career will move at a faster pace.

Additionally, when you can take your understanding of the interest of the organization and explain to others and help them, you can make a huge difference. When you improve the results and yield of others, you add great benefit to them, the company, and yourself. You are a multiplier of value when you help others be more effective.

The opportunity is ready and waiting, and the choice is yours. Strong leaders are hard to find, and companies are searching for them.

Close

You stand at a crossroads in your career. You have gained understanding that many never manage to wrap their minds around. These lessons have provided insight and knowledge that you can leverage to be more effective and raise the value you bring to your clients, your peers, your business, yourself, and your loved ones.

But knowledge alone will not deliver the results you seek. Without your commitment and a little bit of effort, nothing will change.

Remember, you are not defined by yesterday's events—what matters are today's choices and actions. You are in control, so you get to choose how far you'll go and what level of success you'll attain. You decide how much of your potential you will maximize, how much value you will contribute, and ultimately where your life and career will lead.

It all starts with a commitment to yourself. Who are you? What are you going to be? Will you push yourself outside your comfort zone and embrace growth? Are you going to be just a good employee, or are you going to magnify your impact and become a key contributor to the success of the company?

You are not at the mercy of fate—you decide your next steps. Every interaction, every task, every choice you make has an impact. How far do you want to go?

If you need support clarifying these questions on your journey, reach out to me at www.up-skill.com.

I wish you the best in all.

Acknowledgments

At the close of this book, I come to at the end of a task I truthfully questioned if I'd ever complete. Funny enough, you may be feeling the same way as you struggle to finish it.

I am overwhelmed and humbled as I remember the myriad people who have encouraged my journey. When I pause and reflect over the many mentors, leaders, exemplars, coworkers, and friends who have made a difference in my life, I am full of gratitude.

I am grateful for the lessons that I have been exposed to, for the challenge to stretch I was given, and for those who believed and invested in me.

I wish there was a word that expressed that feeling when your throat closes, your eyes tear up, and your chest squeezes your soul, for I would invoke such a word to describe the emotion that comes over me as I think of all those who helped me on my path.

"Thank you" is significantly inadequate.

To the editors and team at Author Bridge, thank you for your wisdom, expertise, and encouragement.

To my family and loved ones, thank you for your

patience, as you have undoubtedly put up with me and my "Dansplaining" for the longest time. I implore your continued patience as I tilt at the next windmill. I am sorry, but to surrender is not in me. I will forever be chasing the horizon where the crimson sunset collides with the teal blue.

With humility and gratitude.

About the Author

Dan Adams is a tenured expert on technology services and service organizations. He is the founder and chairman of New England Network Solutions, Inc. (NENS) a successful TSP in the Greater Boston market who over the past 30 years has served hundreds of companies. Under his leadership, NENS has received numerous industry awards, even being named one of Inc. Magazine's Best Workplaces multiple times.

In 2022, Dan founded Up-Skill to help service professionals elevate their performance and increase their value to their companies and clients and in the process, boost their careers. Drawing on over three decades of experience and a deep love of metaphor, Dan helps team members reduce game changing mistakes and focus on more value contributing actions.

In his free time, Dan enjoys international travel, plays several instruments, and smokes a killer brisket.

Connect with us!

Do you want to magnify the value you get from this book?

Get your FREE Manager Study Guide (MSG) now.

The MSG guides leaders through the seven lessons so you can:

- Improve retention of key points
- Focus on unique areas in your company
- Align expectations and facilitate mutual understanding
- Secure commitment to improvement
- Fast-track implementation of game-changing behaviors

Get your free MSG at www.up-skill.com/msg.html

Here's what some of our readers have to say:

"The guide really helped my team get on the same page..."

"...the book is great, but putting it into action is even better..."

"...I have technical people talking in business terms and thinking ahead!"

Want to dominate the competition, win consistently, and create a dynasty?

People want to contribute and be part of a winning team, but this is not second nature to most. People need to be shown how.

Leaders know that training their people is essential, but rarely do they have the time to create the content, deliver it, and follow up enough to make a real difference. Training is sporadic at best.

Up-Skill also offers training for all members of your team to ensure that, at a minimum, they understand the rules of the game, how their actions impact the team, and how to strategically contribute to winning.

Courses include:

Service Professional

Everyone in the industry knows that to succeed in services you need more than technical skills—you also need soft skills. But what exactly *are* soft skills? Everyone has their own interpretation and how do you even focus on such a nebulous subject?

Our service professional course breaks down the skill areas of success into the Service Success Quadrants™. This delivers crystal-clear understanding and measurable, actionable targets to develop career success for all your service professionals and raise company performance. Do not leave it up to chance or a hope in the existence of common sense. This course teaches everyone the minimums of being a successful service professional.

Core Concepts

This course will increase the number of people on your team who understand the basics for business success in services. You have people making decisions, let's make sure they understand the scope of their impact so they can make the best decisions.

From business-model basics to creating value and margin, this course educates your employees on the core concepts for IT services firms so they can be situationally aware and make better business decisions. Wouldn't it be nice if your influencers (no, we're not talking about social media) promoted good habits instead of bad ones?

LeaderBuilder™

Companies rise to the level of their leadership. LeaderBuilder™ is a unique, patent-pending system that expedites the growth of leadership skills throughout the organization, eliminating years of slower progress.

Average leadership is a bottleneck to employee and company success. Leverage the power of monthly focused lessons and practice to bolster the business acumen of your managers and raise performance across the entire service team.

Join us at www.up-skill.com today to put success on overdrive.

Made in the USA
Monee, IL
20 January 2024

51677055R00085